How to Design Wastewater Systems for Local Conditions in Developing Countries

How to Design Wastewater Systems for Local Conditions in Developing Countries

David M. Robbins and Grant C. Ligon

PUBLISHING

INTERNATIONAL

Published by **IWA Publishing**
Alliance House
12 Caxton Street
London SW1H 0QS, UK
Telephone: +44 (0)20 7654 5500
Fax: +44 (0)20 7654 5555
Email: publications@iwap.co.uk
Web: www.iwapublishing.com

First published 2014

British Library Cataloguing in Publication Data
A CIP catalogue record for this book is available from the British Library

ISBN: 9781780404769 (Paperback)
ISBN: 9781780404776 (eBook)

RTI International is an independent, nonprofit institute that provides research, development and technical services to government and commercial clients worldwide. Our mission is to improve the human condition by turning knowledge into practice. RTI International is a trade name of Research Triangle Institute.

Contents

Dedication

This manual is dedicated to Lisa Lumbao and the staff of the Philippine Sanitation Alliance for their adherence to the principle that everyone has the right to affordable sanitation, and that this can be achieved through the sharing of knowledge, best practices, and lessons learned.

The staff of the Philippine Sanitation Alliance, through which the guiding principles of this manual were conceived. From left to right: Milag Ballesteros, Rhodora Sanidad, Jenny Palmera, Jay Tecson, Lisa Lumbao, David Robbins, Lou Capule, and Ritchelle Urgelles. Not pictured: Nene Narvaez, Andrea Echavez and Lito Santos.

Acknowledgements

The success of this type of manual is based on how well it provides useful, applicable, and accurate information, and to what extent it motivates people into action. To achieve this, the authors have reached out to a number of collaborative individuals and organizations that were willing to share their knowledge, information, case studies and best practices in the hopes of creating a simplified approach to a complex problem. Their collective knowledge is represented throughout as the foundation of this manual. The authors sincerely thank the organizations and individuals listed below for their contribution to this manual:

- The Department of Water and Sanitation in Developing Countries at the *Swiss Federal Institute of Aquatic Sciences and Technology (Eawag/Sandec)*, Dübendorf, Switzerland, creators of 'The Compendium of Sanitation Systems and Technologies', referred to through the manual as 'Compendium'. The 2014 (2nd revised) edition, authored by Elizabeth Tilley, Lukas Ulrich, Christoph Lüthi, Philippe Reymond, and Chris Zurbrügg, is cited frequently and can be accessed online: http://www.eawag.ch/forschung/sandec/publikationen/compendium_e/index_EN
- David Lindbo and the *Consortium of Institutes for Decentralized Wastewater Treatment*, developers of the Decentralized Wastewater Systems Glossary (2009) and Installation of Wastewater Treatment Systems 2009 manual, of which 'Chapter 4: Soil and Site Concepts for Installers' written by David Lindbo was frequently used. Through their kind cooperation, they have provided text and images for this manual, which are either referred to as (CIDWT, 2009), (Lindbo, 2009) or (Lindbo, 2013), the latter of which is from personal communication. These and with other excellent resources can be accessed at: http://www.onsiteconsortium.org.
- PhD students Matthew Verbyla and Erin Symonds at the University of South Florida (USF), Mercedes Iriarte at the Universidad Mayor de San Simon in Bolivia, and others within the international, collaborative 'Reclaim' *Initiative* led by USF (http://usf-reclaim.org/).
- *Institute for Global Environmental Strategies* (IGES) with special thanks to Shom Teoh for sharing her work on community-based grease management programs in Thailand.
- *North Carolina State University*, Raleigh, North Carolina, USA, Department of Civil and Environmental Engineering and Department of Soil Science. Special thanks to Dr. Mike Hoover, Cary, North Carolina, USA, who generously provided guidance and many graphics, referred to as (Hoover, 2013).
- *North Carolina Division of Public Health*, with special thanks to Steven Berkowitz for his kind review and many suggestions for improving this manual.
- *Coconino County Environmental Services*, Flagstaff, Arizona, USA. Special thanks to Joelle Wirth, manager of the on-site wastewater management program, for her insights into soils and site-evaluation techniques.
- The *International Water Association* (IWA) (http://www.iwahq.org/1nb/home.html). They have been key in the development of this publication, not only as publishers, but as a major organization in promoting decentralized wastewater management systems through global conferences, workshops, and collaborations.
- Special thanks to Vince Delector, water and sanitation engineer for *Habitat for Humanity*; Aurélien Vétil of *One Acre Fund*; Heather Durand of *Duke University*; and Julian Doczi of the *Overseas Development Institute* for their reviews, input, images, and advice on practical applicability.

Images and graphics were cited from the latter two contributors either as (Durand, 2013) or (Doczi, 2010, 2012).

Many other individuals and groups assisted with the preparation and review of this manual. The authors sincerely appreciate the widespread enthusiasm for this project, and the level of interest in improving sanitation through decentralized wastewater management.

Foreword

David Lindbo, professor and extension specialist, Department of Soil Science, NC State University.

Clean water is something we all want and need to survive. Being able to supply a population with clean water has often focused on the treatment of water to bring it to safe drinking water standards. This is indeed a laudable goal, but it can be hampered by the quality of the water being treated. In more rural areas where the population relies on shallow wells or even surface water, large-scale potable water treatment may not be always be feasible. So what can be done?

To answer this question, consider why water in many rural and urban fringe areas is polluted. Over the past few decades, many international and national organizations have identified poor sanitation as a source for drinking water contamination. This contamination often comes from the total lack of wastewater treatment facilities or poor maintenance of those that exist. In rural areas, there is often no infrastructure for centralized water or wastewater treatment. The lack of infrastructure is not due to a lack of desire, but most often due to the prohibitive cost of central systems that are more common in urban areas where the economies of scale can make these systems affordable. So what are these infrastructure-poor areas to do?

The solution, applicable in many areas of the developing world, is age-old and has been adopted effectively in much of North America and Europe. *Use the soil to treat the wastewater!* We are not talking about a simple outhouse, but a system that is scientifically designed to not only disperse the wastewater but also to treat it in the soil. The soil can treat the wastewater effectively before it reaches groundwater and eventually surface water, where it can be used for drinking, bathing, irrigation, and a whole host of other uses. By designing a system that relies on the soil, the water resource is protected, and the costs of drinking water treatment can be reduced.

Soil-based technologies are not the only systems applicable in developing country settings. However, where land is available and soils are adequate, such systems are not only cost-effective, but often are the most protective of public health. This manual is a positive step in moving towards a sustainable approach to developing wastewater infrastructure by providing the step-by-step procedures for choosing the most appropriate technologies for particular projects and locations, including soil-based treatment and dispersal systems.

This manual focuses on the design of wastewater systems with the idea that one size does not fit all, but rather, that there are principles of design common to all systems. It takes a step-by-step approach to design so that an individual, community, or region can choose what is best for its wastewater management needs. The process involves four general tasks – source characterization, site evaluation, technology identification, and technology selection. The use of checklists throughout the process makes the design process logical and straightforward. It does not imply reliance on any one technology to solve all issues and recognizes there are many ways to achieve similar results. The acknowledgement that soils may be a potential treatment medium for particular sites enhances this manual. By understanding the nature of the soils and site conditions, designers can use the guidance provided in the manual to select the most appropriate and cost-effective technologies and ensure they are designed to address any identified limitations. In many cases, this makes wastewater systems more accessible to more people who may not have access to highly technical solutions.

As you use this manual, you will find that the authors have taken a holistic view towards helping you address your wastewater needs. It provides a foundation for how to design a system to your locale and also gives you the understanding as to why certain approaches are chosen. The approach will help you choose and design a system that is right for you and the situation you live in. In the end, you will get the solutions you need for your local conditions rather than solutions that may not fit your local situation.

Quick guide – New project checklist

Use this checklist for investigating key parameters and technologies for new wastewater projects. Find additional information on the pages indicated:

Task	Yes √	No √	More information
Source Characterization			
Wastewater composition and types			Page 12
Flow rate			Page 19
Flow strength			Page 22
Flow variability			Page 24
Nutrient loading			Page 25
Other problematic chemical constituents			Page 26
Chemical and physical considerations			Page 27
Microbiological considerations			Page 28
Sampling and analysis			Page 30
Site Evaluation			
Site plan sketch			Page 38
Test holes			Page 43
Soil horizon properties			Page 44
Water table and other limiting conditions			Page 48
Slope and topography			Page 49
Long Term Acceptance Rate and related concerns			Page 51
Surface water features			Page 52
Land use patterns and conflicts			Page 52
Utilities and related conflicts			Page 53
Site accessibility			Page 57
Reuse opportunities			Page 58
Regulatory requirements			Page 59
Technology identification			
Technology categories			Page 62
User interface and pretreatment technologies			Page 63
Conveyance (sewer) technologies			Page 68
Primary treatment technologies			Page 76
Secondary treatment technologies			Page 81
Tertiary treatment technologies			Page 99
Final discharge or reuse options			Page 101
Level of required treatment			Page 101
Technology selection			
Flow to land availability ratio			Page 108
Technology information comparison			Page 111

Chapter 1

Introduction to decentralized wastewater management systems

This chapter will

- Introduce the subject and importance of decentralized wastewater management systems;
- Introduce the purpose, need and audience for this manual; and
- Present the terms and definitions common to these systems.

Figure 1.1 Mr. Noun Makara (upper right) and his crew manufacture components for latrines outside of Phnom Penh, Cambodia. Utilizing the power of markets is an important tool for scaling up wastewater service delivery, as shown in this capacity building project by IDE Cambodia: www.ide-cambodia.org (Authors).

About Figure 1.1

The photos of Mr. Makara's latrine factory near Phnom Phen, Cambodia (Figure 1.1), illustrate how the power of markets can be used to scale up sanitation. Market-based sanitation means matching the ability to supply products with programs that enhance demand and provide incentives fostered by engaged local governments. The major components are:

(1) ***Demand generation***: *People are generally willing to pay for improved sanitation if they believe it will address their needs, wants, and desires. Understanding what these may be for a population means information must be gathered. Development specialists use tools such as customer surveys, focus group discussions, and interviews with key stakeholders to find out what people really want, what they think, and what they think they need related to sanitation. In some instances, their overriding desire is to improve the health of their children. In other areas, it might be the convenience of having indoor plumbing or the status of having a flush toilet. When sanitation technologies are promoted to address the needs, wants, and desires of people treated as customers, they will be more willing to pay for services. Also, it is important to note that to achieve sustainability, the sanitation improvement must fulfill the promises that are made, or people will find other uses for their money.*

(2) ***The sanitation value chain***: *The value chain of decentralized wastewater management describes how technologies are manufactured, sold, delivered, installed, operated, and maintained. To sustain the value chain, products should be manufactured using local materials and labor, and service providers should be trained through strong capacity-building exercises. When demand is generated (as described above), the value chain produces sales, profits, jobs, and economic development.*

(3) ***The enabling environment***: *This is the combination of incentives and the regulatory framework that promotes rather than inhibits participation in sanitation improvement programs. Good programs include incentives that offer tangible benefits to those that comply, as well as enforcement provisions for those that do not. It is also the way local governments interact with citizens and businesses. Frameworks that speed up and simplify a fair and transparent regulatory process can be powerful tools local governments can use to scale up sanitation using decentralized wastewater management.*

Achieving sanitation improvements at scale requires evidence-based promotions campaigns, a robust value chain, and an enabling environment that works. When all three of these elements are present together, improving sanitation can be a strong force for major social and environmental change.

1.1 BACKGROUND

In the past several decades, many organizations have described the lack of basic sanitation for much of the world's population. Estimates put the number of unserved people at more than two billion, mostly in South and East Asia and sub-Saharan Africa. In those regions there are 45 countries where sanitation coverage is less than 50% (Figure 1.2).

While the estimates and definitions of access to basic (often referred to as 'improved') sanitation have been challenged, it is widely agreed that this problem represents a crisis given the strength of the link between poor sanitation and disease and death, especially in children. Poor sanitation results in a significant yet preventable health and financial burden on individuals, families, communities, and countries. The economic burden is quite significant. For example, in India it was estimated that in 2006 the cost of poor sanitation amounted to 6.4 percent of the country's gross domestic product (World Bank Water and Sanitation Program, 2010).

To meet the widespread need for wastewater treatment, centralized sewerage systems have been implemented as the norm for large, medium, and even small sized cities and municipalities in developed countries. However, it is unlikely that centralized wastewater treatment systems will be used in developing country settings except in the most densely populated urban centers. The rest of the population, as well as the commercial, industrial, public, and other facilities lacking proper sanitation systems will need to rely on site-specific solutions using on-site and decentralized wastewater treatment, dispersal, and reuse technologies. The benefits of smaller, decentralized wastewater management systems are many, as they use simpler technologies, allow for more cost-effective reuse of treated effluent on or near the site where the wastewater was generated, are generally simpler to install and maintain, and can be a force for economic development. Indeed, they not only improve sanitation and environmental health, but they can also lead to job creation along the entire value chain, from equipment manufacturers to system designers, installers, and operators.

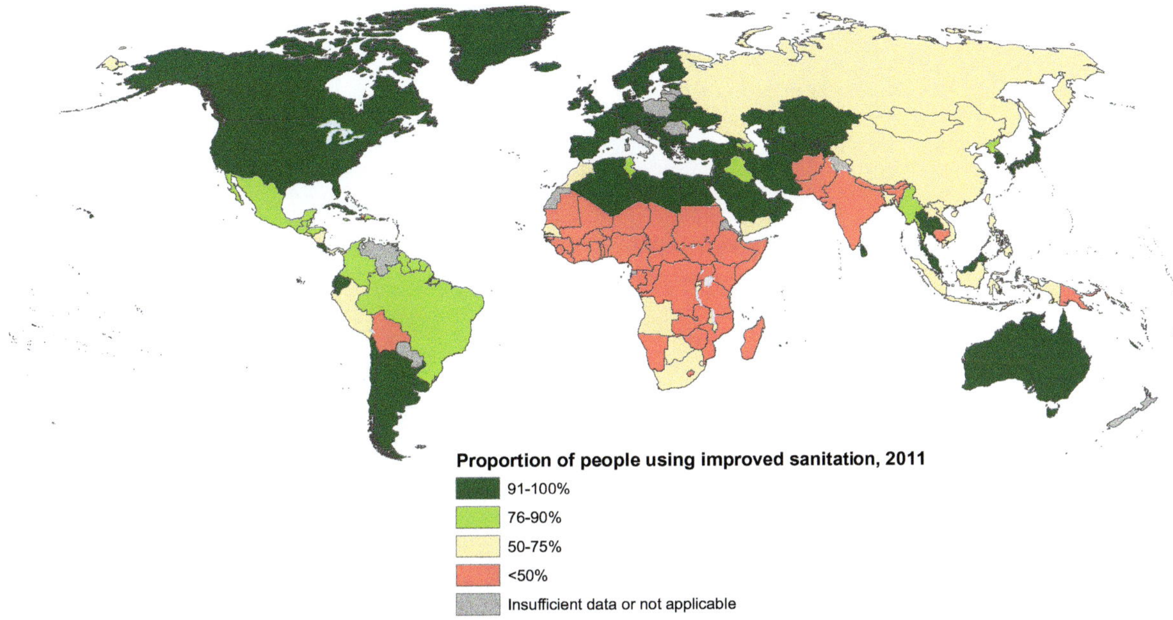

Figure 1.2 Global overview of country-wide estimates of the percentage of countries' populations served by basic sanitation facilities (WHO/UNICEF, 2013).

1.1.1 Combining Technologies to Form Systems

Decentralized wastewater management systems, like their centralized counterparts, use a variety of technologies that work together to achieve many goals. The major goals are to remove sewage away from humans, reduce the pollution from the collected wastewater, and safely disperse the treated effluent, or reuse it and other residuals for beneficial purposes. These are accomplished by linking appropriate technologies together to form systems that meet the specific needs of the end-users. These systems, which vary widely in both complexity and cost, can include:

- toilets, or the human interface that includes individual toilets, latrines, and community-based toilet blocks that serve as the first point of collection for human waste and help to remove it from potential human contact;
- sewers and wastewater collection technologies that collect wastewater from the interface and deliver it to locations for treatment;
- treatment technologies that treat wastewater to a level that is 'safe' for discharge, reuse, or dispersal; and
- reuse and recycling, which represent the technologies that allow for the safe reuse or recycling of the treated effluent or other residuals from the wastewater treatment process, such as biosolids and biogas.

Selecting the most appropriate technologies requires knowledge of the wastewater source and an understanding of the site conditions where the wastewater will be managed. Knowing this information will help service providers to optimize their systems as well as to avoid costly and time-consuming mistakes that could come from improper technology selection. Given their importance, the procedures for conducting wastewater source characterizations and site evaluations in support of decentralized wastewater management systems decision-making are the focus of this manual.

1.1.2 Evolution of the Concept

In recent years, many organizations have used different terminology to describe the concepts of on-site and decentralized wastewater management. The term 'on-site wastewater treatment systems' has been used widely in the United States for many years to describe septic systems and leach fields, or other wastewater systems that manage the wastewater on the site where it was generated. The United States Environmental Protection Agency (US EPA) popularized this term in their 'Onsite Wastewater Treatment Systems Manual', first published in 1980 and revised in 2002. Figure 1.3 shows the cover of the 2002 book, which represents its focus on septic tanks and leach systems (colored in white) that serve many types of institutions and housing clusters. This is a good visualization of the decentralized wastewater management concept that forms the basis of this manual about 'How to Design Wastewater Systems for Local Conditions in Developing Countries'.

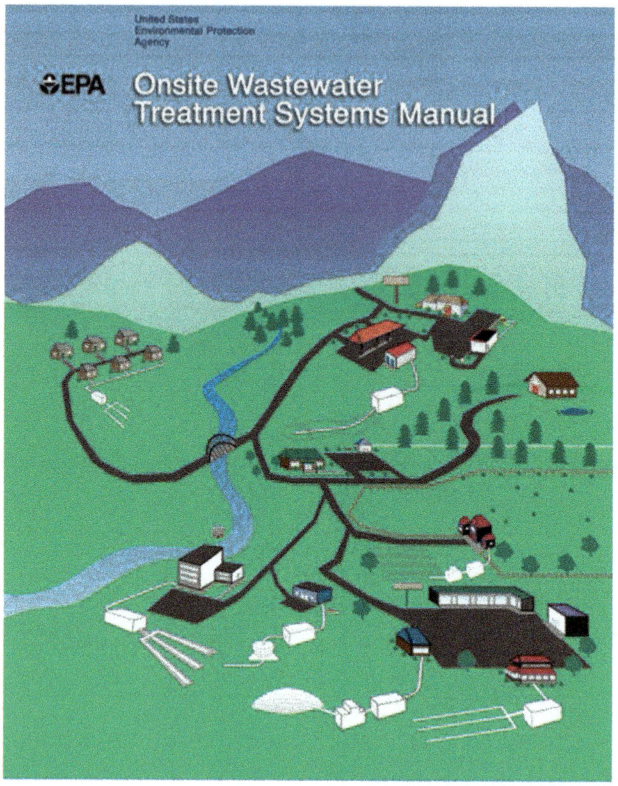

Figure 1.3 Cover page of US EPA 2002 Onsite Wastewater Treatment Systems Manual (National Environmental Services Center at West Virginia University, 2002).

DEWATS, or Decentralized Wastewater Treatment Systems, is an acronym popularized by the Bremen Overseas Research and Development Association (BORDA) in the 1990s. BORDA considers DEWATS to be the various technologies for primary and secondary wastewater treatment that have relatively low-cost operation and maintenance requirements and are most suitable for low-income and sub-tropical regions (Sasse, 1998). The major example of their approach is their DEWATS model, which uses a biogas digester and anaerobic baffled reactor for primary treatment and a constructed wetland for secondary treatment.

The International Water Association (IWA) is also advancing the practice of decentralized wastewater management in developing countries and has recently introduced the concept of *wastewater management* as a more holistic approach. In their Conference on Decentralized Wastewater Management in Asia 2012, which occurred with the support of BORDA in Nagpur, India, that concept was widely promoted by including topics of wastewater collection, treatment, reuse, and recycling of the treated effluent, rather than just wastewater treatment and dispersal. Wastewater management also includes planning, financing, promotion, and regulation of all aspects of wastewater, from its generation to its ultimate reuse or final dispersal. For these reasons, the authors of this manual use the acronym DWM, which stands for *decentralized wastewater management*, and the related term DWMS, which is used in this manual to mean *decentralized wastewater management system(s)*.

1.1.2.1 A Note on Scale

Defining DWMS is somewhat difficult because decentralization is a concept of scale. Unlike centralized systems that collect and manage sewage at one location, decentralized systems manage the sewage closer to where it is generated. Under this definition, a subdivision with 100 homes that uses individual septic tanks and leach fields for each house would be using a decentralized approach to wastewater management. But would a common system that manages the wastewater from all 100 homes be considered a DWMS? For the purposes of this manual, the answer is 'yes', as the management is closer to the sources than if it involved linkage to a single system that served an entire city. But what if 1,000 homes in different areas combined their wastewater flows into a common system? Would these still be considered DWMS? The answer is that for the purposes of this manual, it does not really matter. This manual's goal is to provide useful information on the step-by-step approach of source characterization and site evaluation to make informed technology decisions. When properly trained, local government officials, service providers, or private developers can utilize the process, which is applicable to wastewater projects of all size.

1.2 THE PURPOSE OF THIS MANUAL

People in developing regions of the world generally want improved sanitation and healthy children, and increasingly are seeing the link between improving wastewater management and achieving these goals. Just as important, they often have the ability to pay for wastewater management and the desire to do so. Also, there is often sufficient land available to install systems, and locally available materials, labor, and resources to construct and maintain them. The missing ingredient is the 'how to', meaning how to move from wanting proper wastewater management to implementing a system.

This manual aims to provide the needed information that will help communities that want to improve sanitation utilize the procedures and practices that can make DWMS a viable community-based solution. It details a process for making informed technology decisions based on the specific nature of the development project. This manual can serve as a handbook, field guide, text book, and resource for those with the desire to implement DWMS. In particular, it can serve as an important part of necessary long-term capacity-building programs involving activities such as training, 'twinning' where more advanced utilities (donors) join with less advanced utilities (recipients) for sharing of best practices, job sharing, mentoring, and apprenticeships. This manual's concepts serve as effective methods for teaching and certifying practitioners and may help institutions of higher education find donor funding for train-the-trainer programs or establish degrees, certificates, or in-service educational products.

Furthermore, this manual addresses the need for better methods of wastewater technology selection, which is achieved through promoting a 'context-specific' approach to DWMS technology selection. The process identifies which technologies are most appropriate given the source characteristics and site and other resource limitations. Once the data gathering process is complete, appropriate technologies may be selected to meet the specific needs of the owner.

This manual includes check lists and step-by-step procedures to help users track the data gathering activities, and case studies to illustrate practical applications. Tools, forms, and Internet links to additional information help the user organize the collected data so that it may be used to base sound decisions for positive results. Case studies highlight how other project implementers in similar settings have collected and used relevant information to address their wastewater management needs. These case studies and practical applications are presented to inspire the reader to action by showing that these tasks can be accomplished with available resources.

Finally, site evaluations and wastewater system designs are services that can be performed by the private sector and in many instances, local government. It is hoped that this manual serves as useful guidance to the practitioners doing this work, and that it simplifies tasks and improves results.

1.3 THE AUDIENCE FOR THIS MANUAL

This manual is intended to benefit three major audiences:

(1) Private-sector DWM service providers;
(2) Local and national government regulatory officials tasked with DWM oversight; and
(3) Development specialists and researchers, such as those within NGOs and civil society groups working to promote and scale up sanitation improvement.

Building the capacity of private sector DWM service providers (Figure 1.4) is critical in scaling up sanitation improvement. This core group of individuals and business owners are the first audience for this manual, and includes:

- engineers and service providers that will be tasked with planning the DWMS of the present and the future;
- contractors that will be required to install the system components using best installation practices;
- septic tank desludgers and general maintenance staff who will be paid to operate and maintain systems;
- plumbers and electricians who will be called upon to help the contractors properly install systems; and
- technology providers that must ensure that the components they sell to end-users are appropriate given the specific needs and conditions of the project.

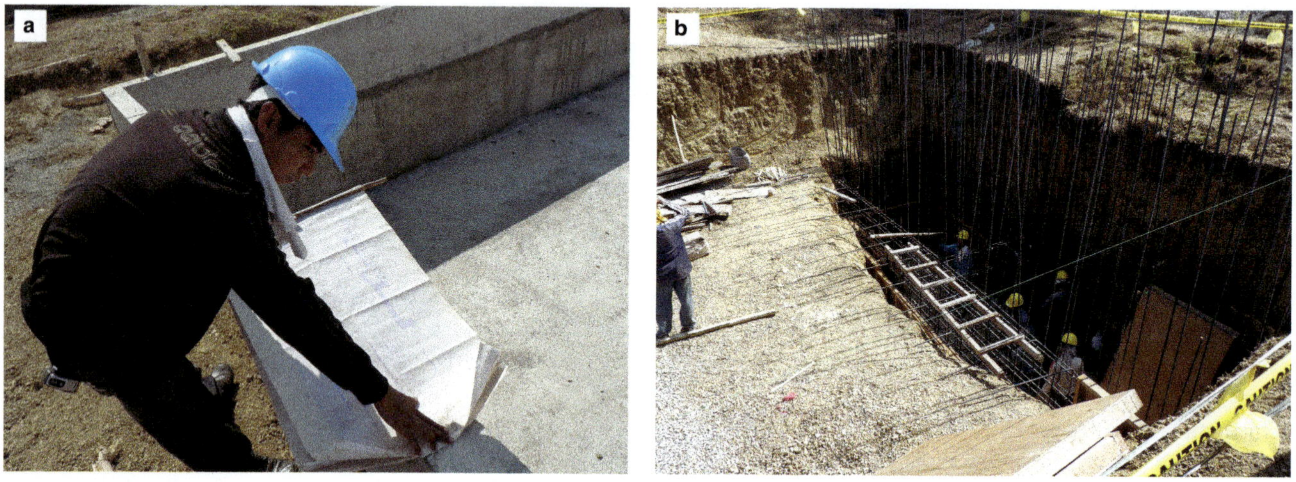

Figure 1.4 Engineer (a) and contractors (b) building a septage treatment tank in San Fernando City, Philippines (Authors).

The second audience for this manual is local government officials who are responsible for wastewater management in their communities. These officials may manage local sanitation plans, enforce local wastewater laws, or oversee planning and installation of DWMS. In many instances, local governments have yet to implement procedures for DWMS, but many are thinking about it. Agencies that apply the information in this manual to develop their DWMS programs will be more likely to see the real and sustainable benefits for their community.

The third audience for this manual is development workers who promote sanitation through the power of DWMS. Development projects are often the first sanitation interventions in a community or region. Development programs funded by international donors may, in many cases, begin by working with local public and private stakeholders to create pilot projects. Local pilot projects that follow a step-by-step approach and create systems that are appropriate for local conditions can be a model for other projects in the community or region.

1.4 THE DWMS DEVELOPMENT MODEL

The DWMS Development Model (Figure 1.5) provides the basis for the step-by-step process for making informed technology decisions. It also suggests a method for integrating wastewater decision-making into the local government's existing compliance structure. Each step in the process could be associated with officials' physical inspection, with plan review by trained specialists, or through self-compliance and reporting programs. The premise of the model is as follows:

> *In order to determine the most appropriate technology choices of a DWMS for managing a specific wastewater flow, it is first necessary to properly characterize the wastewater source. Then, the site where the wastewater will be managed must be evaluated. These two steps must follow the principles stated below.*

- The source must be characterized to determine the:
  - volume of wastewater to be treated, otherwise known as the hydraulic loading or *design flow*;
  - concentration of organic matter and other contaminants in the wastewater, otherwise known as *wastewater strength*;
  - variability of the source conditions as they change over time; and
  - other constituents or parameters that define the wastewater source.
- The site must be evaluated. This involves determining the:
  - *conditions* of the site, including soils properties, slope and topography, the presence of groundwater and surface water, land use, utility issues, accessibility, reuse and regulatory concerns, and other aspects that may restrict the site's capacity to receive and process wastewater. It also suggests the range of DWMS technologies appropriate for the site.

The DWMS Development Model is represented in Figure 1.5 below.

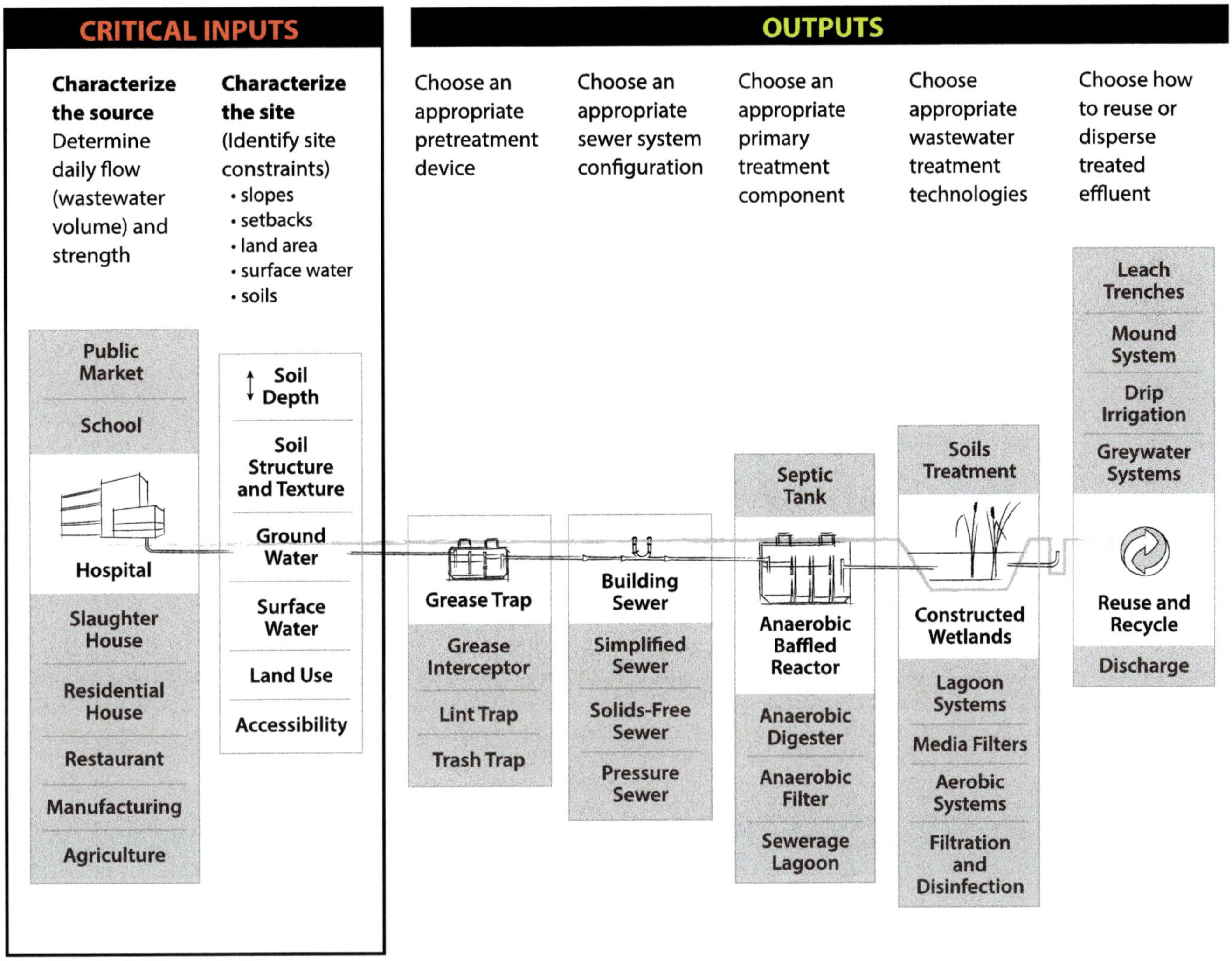

Figure 1.5 The DWMS Development Model.

Fortunately, performing source characterizations and site evaluations are well within the means of local service providers and can often be accomplished with minimal cost and in a short period of time. The information gathered will inform the user about the necessity for pretreatment units, the most appropriate sewer configurations, and which treatment and dispersal or reuse technologies might be combined to achieve the desired end results.

1.5 HOW TO USE PRINT AND WEB VERSIONS OF THIS MANUAL

This manual includes text, step-by-step guidance, checklists, case studies, practical applications, and references. This manual is also available on the IWA's waterwiki website, the 'Information Resource & Hub for the Global Water Community' at http://www.iwawaterwiki.org/xwiki/bin/view/Main/WebHome.

On this manual's electronic version, users will find active links to information found on the waterwiki website. All of the forms and checklists found in the manual may be downloaded and printed. The electronic version provides links for greater detail on the case studies, and other references and interactive toolkits can be explored and downloaded. To help the reader navigate this information, the icons shown below will direct readers to external resources and key points. In the electronic version of this manual, click the icon to link directly to the information. The icons used in this manual are as follows:

Checklists: These help the user keep track of their progress in important steps in the process.

Case studies: These are an important way of sharing information on how others have addressed similar issues.

Toolkits: These computerized spreadsheets are useful in certain aspects of source characterization, site evaluation, and technology comparison and unit sizing. They should be considered as guidance only, and not as a substitution for actual engineering, as the design should be performed by trained experts with specific knowledge of the site.

Web resources: These resources relate to the procedures, technologies, and general information on topics discussed within the manual. This icon implies that the source documents are either provided on the waterwiki website or provided hyperlinks.

Best practices: These are lessons learned from other DWM practitioners as they have worked to overcome difficult challenges of certain DWMS. These best practices have been thoroughly developed and reviewed and deserve special attention. This icon indicates that the source documents are either on the waterwiki website or provided hyperlinks.

Case studies are presented in tan text boxes.

Best practices are presented in blue text boxes.

1.6 DEFINITIONS AND TERMS

The following terms and definitions are adapted in part from the glossary of the Consortium of Institutes for Decentralized Wastewater Treatment (2009).

Aerobic Treatment Unit (ATU) – 1. Treatment component that utilizes oxygen to degrade or decompose wastewater, with or without mechanical means; 2. Term traditionally used to describe proprietary devices that use direct introduction of air into wastewater by mechanical means to maintain aerobic conditions within the treatment component.

Anaerobic baffled reactor – An improved septic tank that uses multiple compartments through which the wastewater is forced to flow upwards through the sludge blanket to trap solids, where they are retained and reduced through anaerobic digestion.

Anaerobic digestion – The degradation of concentrated wastewater solids by anaerobic bacteria. These break down the organic material into inert solids, water, carbon dioxide, and methane, which can serve as an energy source.

Baffle – In a septic tank, an inlet baffle is a pipe, tee, or wall segment at or near the inlet pipe of a tank which is designed to slow down and direct influent flow below the wastewater surface, isolate scum from the inlet pipe, and allow ventilation. An outlet baffle is a pipe tee or wall segment at or near the outlet pipe of a tank designed to collect flow from the clear zone, isolate scum from the outlet pipe, and allow ventilation.

Biosolids – Dewatered, primarily nutrient-rich organic material generated as a byproduct of biological wastewater treatment processes that can be recycled (such as for use as a soil amendment).

Biochemical oxygen demand (BOD) – Quantitative measure of the amount of oxygen required for the biochemical oxidation of carbonaceous material in wastewater; typically measured after a 5-day period.

Carbon to nitrogen (C:N) ratio – A ratio of the mass of carbon to the mass of nitrogen in a substance. Achieving proper balance by mixing various combinations of plant and animal waste can achieve 'ideal' C:N ratios that maximize microbial decomposition in compost piles, and methane production in anaerobic digesters.

Chemical oxygen demand (COD) – Amount of the oxygen required for degradation of total matter in wastewater that can be oxidized by a very strong chemical oxidant; typically measured by a standard test using dichromic acid as the oxidant.

Combined sewer – A sewer intended to receive both wastewater and storm water.

Coliform bacteria – Group of bacteria that constitute most of the intestinal flora of warm-blooded animals (including the genera *Klebsiella sp., Enterobacter sp., Citrobacter sp.*, or *Escherichia sp.*) and are used as water pollution indicator organisms.

Composite sample – Combined individual samples collected from the same point at different times; samples may be of equal volume or may be proportional to the flow at time of sampling.

Daily Design Flow – Estimated volume of wastewater for a representative 24-hour period; parameter used to size systems.

Desludging – The process of cleaning or removing the accumulated sludge or septage from a latrine, septic tank or wastewater treatment facility.

Domestic sewage – Wastewater of residential strength, or effluent from a septic tank or other primary treatment device with a BOD5 less than or equal to 170 mg/L, total suspended solids (TSS) less than or equal to 60 mg/L, and fats, oils, and grease less than or equal to 25 mg/L. It may include wastewater from commercial or food service sources, but not industrial wastes.

Dry weather flow – The flow of wastewater in a combined sewer during dry weather. Such flow consists mainly of wastewater with no storm water included.

Excreta – Human waste composed of urine and feces.

Effluent – Liquid flowing out of a component or device (the opposite of influent).

Facultative – Having the ability to live with or without free oxygen.

Grab sample – Individual sample collected at a particular time and location.

Helminths – Parasitic worm-like organisms that feed off living hosts and produce eggs that persist in wastewater effluent and septage sludge unless properly treated.

Impermeable – Permitting little or no passage of fluid through pores.

Lagoon – Constructed basin lined with either very low permeability soils, or a synthetic material surrounded with berms. Wastewater is treated via physical, chemical, and biological processes.

Methane (CH_4) – A colorless, odorless, and flammable gas present in natural gas and formed by the anaerobic decomposition of organic matter. In wastewater treatment, methane can be generated by the anaerobic digestion process as a byproduct which can be used for cooking, heating or conversion to electricity.

Ped – In soils, refers to individual aggregates separated from each other by voids or natural areas of weakness.

Pretreatment – Any component or combination of components that provides treatment of wastewater prior to conveyance to a further treatment and dispersal components or reuse. Often, this treatment is designed to meet primary, secondary, tertiary, and/or disinfection treatment standards and may include grease traps for use in restaurants, lint traps for use in laundry washing, and trash traps for use in public markets or where open sewers are prevalent.

Primary treatment – Physical treatment processes involving removal of particles, typically by settling and flotation with or without the use of coagulants (for example, using a septic tank or anaerobic baffled reactor).

Operation and Maintenance (O&M) – The routine and emergency repairs and service that is required to keep wastewater systems functioning properly, which should be performed by licensed or trained personnel using best practices.

Quaternary ammonia disinfectants – Solutions that contain nitrogen that are used to kill microorganisms, commonly used in hospitals, restaurants and homes; excessive use of these can add significantly to the concentration of nutrients in wastewater.

Scum – Layer of floating material on a liquid surface which may include fats, oil, and grease (FOG) or other solid wastes.

Secondary treatment – Biological and chemical treatment processes designed to remove organic matter.

Septage – Liquid and solid residuals (sludge) that accumulate and must ultimately be removed from pretreatment devices, septic tanks, seepage pits, portable toilets, latrines or similar components.

Service provider – A public or private entity engaged in site evaluation, system design, construction, inspection, and operation and maintenance. Includes those that collect, transport, and treat septage and fecal sludge.

Sewage – Untreated wastes consisting of blackwater and graywater from toilets, baths, sinks, lavatories, laundries, and other plumbing fixtures in places of human habitation, employment, or recreation.

Sewer – Otherwise known as sewer collection systems, they are systems of piping, lift stations, and other structures that receive and convey wastewater either by gravity or pressure.

Sludge – Precipitated solid matter with a highly mineralized content produced by domestic wastewater treatment processes.

Wastewater Conversion Factor (WCF) – The percentage of potable water that enters a building that is converted to wastewater, taking into account the volume of water that is utilized for outdoor activities, such as landscape or garden irrigation, or other activities.

1.7 GUIDING CASE STUDY

To illustrate and link the key DWMS concepts and decision-making procedures described in this manual from a *practical perspective*, an actual facility is presented below and used as a model throughout the manual. It shows how a facility owner used the step-by-step process to make informed wastewater technology decisions and reveals the outcome of those decisions.

LORMA Medical Center, San Fernando City, Philippines

The LORMA Medical Center (LORMA) is located on the island of Luzon in the Northern Philippines, in the city of San Fernando in the Province of La Union. It was founded by Dr. and Mrs. Rufino N. Macagba, Sr., who are both physicians. It is a 136-bed facility with 450 workers. Data for this case study were collected through personal observation by the author and interviews with Dr. Macagba in 2009, 2010, and 2012.

The LORMA Medical Facility is located near Carlatan Creek and the national highway. It is a good case study for the following reasons:

- it is typical of many point sources of wastewater in developing countries;
- as there are no plans for a centralized wastewater system to serve this part of the city, a decentralized or onsite wastewater management approach is required;
- LORMA can be considered a difficult site, with a relatively large wastewater flow, a nearby river that is subject to flooding (Carlatan Creek), and not much land available for a DWMS; and
- it has food service and laundry washing activities, along with hazardous and medical waste generation.

Throughout this manual, the LORMA case study will be used to illustrate how the characteristics of the wastewater generated by the hospital activities and the site where the wastewater is managed influenced the technology choices. The reasons for selecting the chosen technologies also will be presented.

REFERENCES

Consortium of Institutes for Decentralized Wastewater Treatment (2009). CIDWT Decentralized Wastewater Treatment Glossary, 2nd edn, 2009. http://www.onsiteconsortium.org/Glossary2009.pdf (accessed 4 April 2013)

Sasse, L. (1998). DEWATS: Decentralized Wastewater Treatment in Developing Countries. http://www.sswm.info/sites/default/files/reference_attachments/SASSE%201998%20DEWATS%20Decentralised%20Wastewater%20Treatment%20in%20Developing%20Countries_0.pdf (accessed 5 October 2012)

National Environmental Services Center at West Virginia University (2002). Cover Graphic for: The US EPA Onsite Wastewater Treatment Manual (EPA625R-00008). http://nepis.epa.gov/Adobe/PDF/30004GXI.pdf (accessed 12 January 2013)

WHO/UNICEF Joint Monitoring Programme for Water Supply and Sanitation (2013). Progress on Sanitation and Drinking-Water – 2013 Update. http://www.childinfo.org/files/JMP2013Final_Eng.pdf (accessed 4 October 2012)

World Bank Water and Sanitation Program (2010). The Economic Impacts of Inadequate Sanitation in India. http://www.wsp.org/sites/wsp.org/files/publications/wsp-esi-india.pdf (accessed 8 July 2012)

Chapter 2

Characterizing the source

This chapter will

- Introduce the various sources and compositions of wastewater common in developing country settings;
- Provide guidance on determining the daily design flow for DWMS;
- Introduce the concepts of wastewater strength and contaminant loading; and
- Present the step-by-step process of obtaining source data through field sampling and laboratory analysis.

Slaughterhouse *Food Service*

Schools *Housing*

Figure 2.1 Some of the many important residential, institutional, and commercial sources of wastewater in all types of communities (Authors).

About Figure 2.1

Figure 2.1 shows some images from daily life around Southeast Asia and Guyana, but they could be anywhere. Slaughterhouses, food service, schools and housing blocks are in just about every community. This chapter will discuss the key point that while these are common facilities that are integral to our daily lives, each has very different wastewaters draining from them. Wastewater from a slaughterhouse for example, and from a home are so vastly different, it is even difficult to make comparisons between them. What is interesting, however, is that while we can encounter widely differing wastewater sources throughout our own community, there is not much difference between common sources, such as public markets or housing blocks that may be located in very different geographic areas. A restaurant in Kathmandu is likely to produce wastewater similar to that coming from a similar restaurant in San Salvador. This fact can be very empowering for service providers that design or install wastewater systems and even owners who want to improve their facility's sanitation.

Case studies in wastewater management (and there are many in this manual) show how people addressed problems that may be similar to your specific needs. With an Internet connection, information and lessons learned from an increasing number of case studies are available with just the click of a mouse. Take a look at some of resources where case studies and lessons learned are freely shared, as well as the references section at the end of each chapter:

- *http://www.iwawaterwiki.org/xwiki/bin/view/Main/WebHome – The 'Wikipedia' of water and sanitation hosted by the International Water Association (the publisher);*
- *http://www.susana.org/ – The Sustainable Sanitation Alliance, with major funding by the Bill & Melinda Gates Foundation and support from hundreds of other partners. Visit their forum, where you can ask an expert questions: http://forum.susana.org/forum/;*
- *http://www.indiawaterportal.org/ – The India Water Portal, by Arghyam (www.arghyam.org), a grant-making institution in India. Many case studies and resources are available that are applicable well beyond Indian borders;*
- *http://www.solutionsforwater.org/ – Solutions for Water is a product of the 6th World Water Forum. Take some time to look around this user-friendly site;*
- *www.sswm.info/ – Sustainable Sanitation and Water Management. An excellent resource with many case studies and tools related directly to the topics in this book. Refer to their resources for practical information and applicability.*

2.1 INTRODUCTION

The first set of inputs for Decentralized Wastewater Management Systems (DWMS) decision-making comes from *characterizing the wastewater source*. Understanding the physical, chemical, and biological aspects of the wastewater and how these may change over time must guide the DWMS design process. Figure 2.2 shows an example of this change over one day from one U.S. house. Wastewater is produced in varying flow rates (the height of each labeled area) and strengths, ranging from weak to strong.

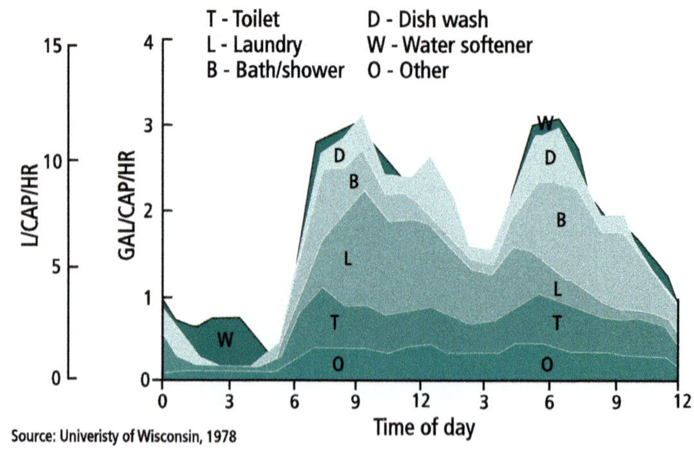

Figure 2.2 Flow pattern for sample U.S. single-family residence (labeled as CAP, for per capita) by hour over a typical one day period (adapted by Onsite Sewage Treatment Program Staff, 2011).

Chemical compositions and other components also vary over one day. These and other important source characteristics can be grouped as follows:

- *physical aspects*: volumetric flow rate, color, temperature, and pH;
- *biological contaminants*: organic matter from human or animal waste, food-related matter, and microbes, some of which are pathogenic; and
- *chemical constituents*: mostly originating from soaps, disinfectants, cleaners, and other products as well as various animal/other wastes rich in nitrogen and phosphorus.

The physical, chemical, and biological aspects of wastewater vary widely from source to source and over time. This chapter focuses on understanding the specific nature of these variables in any wastewater source. Wastewater source characterization can be organized into separate tasks and begins by determining the composition of the wastewater.

2.2 WASTEWATER COMPOSITION

Wastewater from any given source may contain contaminants within the following categories:

- urine and feces from humans or animals (brown waste);
- waste products from processes like food manufacturing, food preparation and utensil washing that can discharge fat, oil, and grease (FOG);
- organic chemicals or contaminants from commercial operations such as dry cleaning;
- inorganic chemicals or contaminants such as lint from laundry washing operations and compounds from metal plating, leather tanning, or printing; and
- water, which may account for the majority of total flow. Wastewater effluent with 1000 mg/L of contaminants (considered high strength) is 99.9 percent water!

2.2.1 Blackwater and Graywater

The composition of wastewater, especially from residential sources, can be thought of as either blackwater or graywater. Given recent focus on more efficient and reuse-oriented management of human waste, programs that separate toilet waste (blackwater) from sink, shower, and laundry washing water (graywater), and reuse that as shown in Figure 2.3 and through other processes, will become more popular. While graywater may have less organic matter and pathogens than separated blackwater or combined sewage, it still must be managed appropriately to avoid disease transmission or water source contamination. In this manual, graywater is considered to be wastewater. Therefore, developing graywater management programs requires the same attention in identifying the specific characteristics of the graywater source as well as the site where it will be treated, dispersed, or reused.

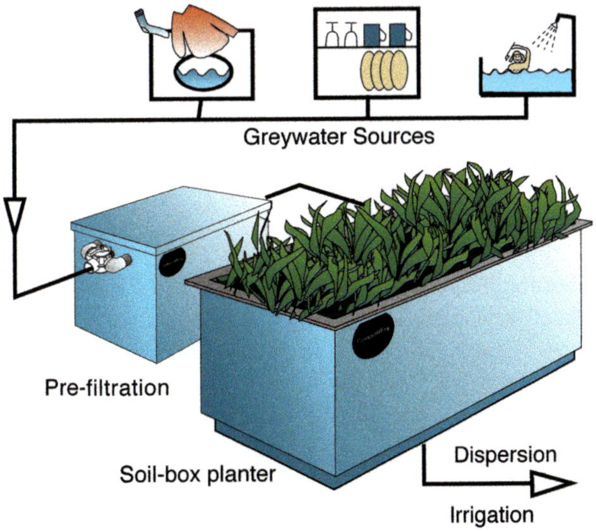

Figure 2.3 Common reuse scenario for graywater from laundry, dishwashing and bathing wastewater flows (Lindstrom, 2000).

In many developing countries, especially in Asia and Africa, the common building practice is to separate the graywater from the blackwater at the source through a separate network of pipes. Blackwater is then sent to septic tanks for primary treatment, and graywater is plumbed directly to open drainage channels for direct discharge with no further treatment (Figure 2.4). Such practices have two impacts:

- blackwater is concentrated in the septic tanks or other treatment units, making it more difficult to treat; and
- graywater is disposed of and not available for reuse, and instead can pond on the ground surface or in a ditch, after which it quickly turns septic, generates odors, and potentially exposes humans to pathogens.

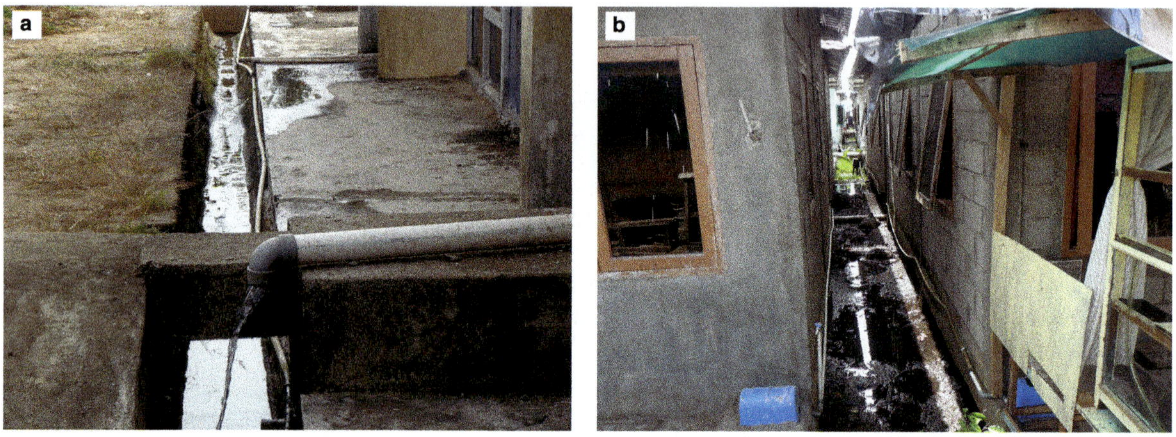

Figure 2.4 (a) and (b) Graywater separated at the source and discharged to open drainage ditch, Palembang, Indonesia (Authors).

When properly treated, graywater can be used in many beneficial activities such as landscape irrigation and toilet flushing. However, wastewater flow, strength, and the ability to derive beneficial reuse of graywater differs greatly based on the source. For example, graywater from laundry or other flows that are heavily contaminated by soaps and disinfectants may prove particularly difficult to treat. This is described in the discussion of the different wastewater sources in the next section.

2.3 WASTEWATER SOURCE CATEGORIES

There are a wide range of wastewater sources typical of communities in developing countries. These may be grouped as follows:

- residential housing, such as single-family houses, multi-family units, and subdivisions;
- public markets;
- slaughterhouses and other food processing operations;
- hospitals and health care facilities;
- restaurants, hotels, and resorts;
- offices, jails, and daily use institutional buildings, as well as those with less-regular use such as churches;
- schools, including those with food service and dormitories;
- commercial operations, such as beauty salons, automobile repair shops, metal fabricators, tanning and leather operations; and
- small communities or commercial centers with combined sources.

Each source category listed above has its own norms related to the activities that generate wastewater. Schools, for example, with their highly variable flows, present much different wastewater challenges than slaughterhouses, which may have more regular flows, but higher concentrations of organic matter. Also, within each category there might be high variability. Consider the differences between a business hotel near a city center and a beach resort with high seasonal flow rate variability. Due to the differences in wastewater from residential, commercial, or institutional operations, each facility must be evaluated by itself within the DWMS planning process.

This evaluation involves looking at the flow rate, strength, composition, and variability of the wastewater. For existing facilities, this may be straightforward if data are available from water meter readings or general water consumption. If that information is not available, it often can be obtained from devices such as flow meters, and through sampling and analysis to determine the specific concentration of various contaminants. For new construction, assessing the nature of the wastewater that *will* be generated often requires relying on estimates and comparisons to similar existing facilities.

The following sections introduce the step-by-step process for conducting source characterization. However, the major common categories of wastewater sources are first presented to help the reader identify general conditions to be aware of when characterizing their source and ultimately using the information gathered to design DWMS.

2.3.1 Residential Housing

DWMS may be used for single-family houses, low-income housing blocks, mid-rise housing developments, and subdivisions (developments of multiple single-family houses). Wastewater from residential housing can contain many types of inorganic and organic chemical compounds that are introduced into the flow through daily activities. Even different single-family homes or housing developments may have very different concentrations of organic matter and suspended solids in their wastewater streams.

Also, wastewater flow rates from housing are strongly related to socioeconomic level, number of residents per unit, geographic location, and the availability and cost of potable water. For example, subdivisions with a continuous water supply (24/7 piped water) will usually have much higher flows than communities that haul water in buckets from shared wells or collect rainwater for domestic use.

2.3.2 Public Markets

Typical public markets in developing countries include wastewater-generating activities that take place at different locations throughout the market (Figure 2.5). These may include:

- stalls where meat processing, trimming, and butchering occur;
- areas where fish are descaled and cleaned;
- sections set aside for vegetable trimming and washing;
- locations where food is prepared and consumed and where utensil washing takes place; and
- public restrooms.

Figure 2.5 Vegetable section at Dong Hoi public market, Vietnam (Authors).

Public markets generate wastewater with higher volumes of organic matter than residential housing. Food preparation stalls often significantly add to the wastewater strength by generating grease, solids, and organic matter. These stalls also produce scraps that should be removed by trash traps. Wastewater systems for markets must be flexible enough to handle the fluctuating flow, from almost no flow when the market is shut down, to peak flows on busy market days. Sewer systems for public markets are often open, allowing storm water, trash, and debris to enter, which make treatment more difficult. A case study of how one designer addressed these conditions for the Alabang public market in The Philippines is presented on the following page.

Case Study: Public market wastewater treatment in the Philippines

Wastewater treatment for public markets in the Philippines is complex due to the relatively high strength of the discharges and variability of flows. The Alabang Public Market, located in Muntinlupa City, is one of the largest public markets in the metropolitan area of Manila, with 1448 stalls and 24 hour-per-day operation. With support from the U.S. Agency for International Development for planning and design, the city built a treatment facility in February 2006 that is functioning well after 8 years. It combines anaerobic and aerobic treatment with filtration using cocopeat media to meet local discharge standards. It also includes a water recycling system that allows reuse of the treated effluent for flushing toilets, watering plants, and street cleaning.

Design Considerations

When the facility was designed, the market had about 4800 active vendors and workers and 4500 daily customers. The average wastewater output was 210 cubic meters of sewage per day with a Biochemical Oxygen Demand (BOD) of around 600 milligrams per liter (mg/L). Suspended solids and fats, oil and grease (FOG) were also quite high. The discharge flowed to the Alabang River which drains into Laguna Lake, with discharge limits of 50 mg/L BOD.

Site-Specific Challenges

Developing an effective wastewater management strategy for the Alabang Public Market required overcoming several site-specific constraints:

- limited available space (160 m^2) for treatment facilities, as the site was used as a parking lot and delivery area;
- a high relative elevation of the outfall in relation to the area designated for treatment (uphill slope); and
- unstable fill upon which to base the facilities (the site was an old landfill).

Due to the very limiting site conditions, a mechanized system was the most practical application. It was fully funded by a small tariff added to the annual stall owner fees, which helped to achieve full cost recovery in 3 years. The facility uses an anaerobic baffled reactor (ABR) for primary treatment, an aerobic treatment system for secondary treatment, and a cocopeat filtration system to further treat the wastewater for reuse, as shown in Figure 2.6.

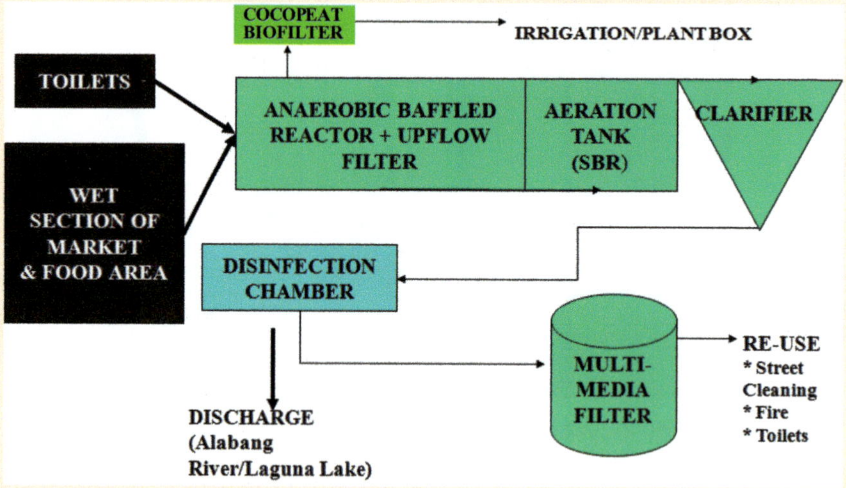

Figure 2.6 Alabang market wastewater treatment plant layout (Authors).

For more information, refer to the article 'Low-Cost Innovative Solutions for Treating Public Market Wastewater in the Philippines: Deploying Hybrid Anaerobic and Aerobic Cocopeat Filtration Systems', by Carlito Santos and David Robbins (2011): http://www.rti.org/pubs/alabangmarketstudy.pdf.

2.3.3 Hospitals and Health Care Facilities

Medical facilities, which may have wastes from patient care, food service, laundry washing, and facility cleaning and disinfection, have complex wastewater streams and treatment requirements. Medical wastes from treatment and operating rooms, hazardous wastes from laboratories, and even radioactive wastes must be segregated and managed in separate off-site waste management facilities, often requiring licenses from central government agencies. Food service and laundry operations need appropriate pretreatment devices, including grease and lint traps. Additionally, the use of quaternary ammonia disinfectants in large quantities may harm beneficial microbial growth in treatment systems, making biological wastewater treatment less effective.

2.3.4 Slaughterhouses

Slaughterhouses generate very high-strength wastewater, with BOD levels sometimes reaching 10 to 20 times (or more) the BOD levels of residential housing. Much of this is due to the blood, hair, FOG, and other organic wastes that flow down the drain. Many slaughterhouses have livestock pens that are periodically flushed, randomly sending more high-strength wastewater into the sewer system. Also, slaughterhouses often use high volumes of disinfectants, which can inhibit the growth of microbes that are beneficial to certain wastewater treatment processes.

While these variable conditions mean that slaughterhouses typically require very robust DWMS, valuable byproducts can be produced from slaughterhouse waste. For example, the collected blood can be used in animal feed, the waste digestion process can be used to generate biogas, and the processed biosolids can be used as a soils amendment.

Slaughterhouses and regulatory authority

In many countries, slaughterhouses are regulated by national bodies that have their own requirements for sanitation and wastewater management. In Kenya, sanitation at slaughterhouses is overseen by the Assistant Undersecretary for Animal Wealth and Agriculture Affairs, under the Ministry of Environment and Water. In the Philippines, due to decentralization, the responsibility for regulating slaughterhouse sanitation falls to local government units. In general, understanding the legal requirements and identifying the regulatory authority is an important first step in achieving sanitation compliance.

2.3.5 Hotels, Resorts, and Restaurants

With moderately high strength wastewater and flow variability, the hospitality industry presents challenges for DWMS. The food service and laundry operations typical of these facilities require more pretreatment units than for residential housing. Daily and seasonal flow variability also may be greater and may depend on the type of the dining or boarding establishment. Table 2.1 shows how the type of food service establishment relates to dining interval and flow rate and variability, based in large part on the extent and method of dishwashing.

Hotels and resorts have different flow variability than restaurants because of peak and low seasons. Their grease and lint management needs are very case-specific. They also have to be concerned about sanitation for different reasons. In many locations, beachside hotels and resorts discharge wastewater with minimal treatment directly into the sea. However, this can be damaging to the establishment if guests smell odors or see discharge from poor wastewater management or if they believe recreational waters are contaminated. For this reason, resort owners are often easy to convince of the benefits of proper wastewater management and are frequently early adopters and supporters of DWMS programs.

Table 2.1 Restaurant classification and related dining intervals and service levels.

Establishment type	Average dining interval	Notes
Fast food	30–45 minutes	Single-serve dishes and utensils reduce dishwashing and flow. Typically greatest amount of grease per customer.
Restaurant	60 minutes	Full dish and utensil washing. Typically greatest amount of wastewater generated per customer.
Leisure dining or supper club	90–120 minutes	Full service with bar likely. Peaks in flows based on weekly or less-frequent events. Generally highest level of flow variability per customer.

2.3.6 Office, Daily use, and Less Regular-use Buildings

Wastewater from office buildings is generally relatively low-strength with regular and predictable flows. Office buildings typically experience high flows during normal working hours. Office wastewater flows are often calculated using the number of workers per shift and the number of shifts in a 24-hour period. Also included in the calculation would be the presence or absence of (1) food service (and if it is full service with utensil washing), (2) showers, and (3) meeting rooms (where outside guests may be coming in). Churches, mosques and other places of worship good examples of less regular-use facilities that also have predictable daily and weekly flows. Their flows greatly contrast those of traditional work facilities, due to the nature of typical worship services that are held throughout the week.

2.3.7 Schools

Schools have highly variable flows and present challenging wastewater management problems. The wastewater discharged during the average school day may be the typical flow, but the system must also be able to handle low flows during weekends and peak flows during special events. Flow equalization methods, which use tanks and pumps to moderate peak flows by averaging them out over time as described in Section 2.6.1, may help address this issue. Also, the systems must be able to restart quickly after lengthy no-flow conditions during summer vacation. Aside from flow concerns, schools with larger grounds may consider reuse programs where the treated and nutrient-rich effluent can be used for landscape irrigation.

2.3.8 Other Commercial Facilities

Many other commercial, agricultural and industrial facilities generate wastewater. Common sources may include the following:

- beauty salons;
- automotive repair and fueling stations;
- car washes;
- tanneries and leather goods manufacturing;
- confined animal feeding operations; and
- dental and jewelry repair services.

Creating DWMS for these facilities requires careful source characterization to properly assess flow volume, strength, variability, and especially unique and potentially very toxic chemical constituents. Comparing these facilities with similar operations and consulting industry-specific trade groups may help to provide useful information for estimating baseline wastewater-related data.

2.3.9 Community Wastewater Systems

Community wastewater management systems receive flows from multiple sources that may include residential and/or commercial discharges, often along with storm water. Wastewater flows may be separated from storm water in sanitary sewers, or comingled with storm water in combined sewers, which is frequently the case in developing country settings. When combined sewers are connected to treatment works, they are often equipped with combined sewer overflows, which are designed to discharge into surface waters during significant storm events when flow levels exceed treatment facilities' design flow. In these cases the diluted wastewater must still meet regulatory standards. Community sewerage projects often use a multi-step process for treating wastewater. This includes pretreatment devices installed at targeted businesses, primary treatment devices such as interceptor or septic tanks at individual homes and buildings, and secondary treatment and disinfection mechanisms to further process the wastewater.

Community-based systems are presented here as they still may fall within the definition of DWMS. The case study provided below introduces an important model of how one city is addressing unregulated wastewater flows from existing residential areas.

Interceptor tank with pump station, San Fernando City, Philippines

An interceptor form of sewage collection (Figure 2.7) serves as the community DWMS for approximately 55 homes and small shops within San Fernando City. Combined storm water and sewage that flows through the community drainage ditch are captured in a concrete tank and pumped to the treatment facility. The wastewater is low strength with high hydraulic loading and is controlled by a storm water overflow. During high flows, diluted wastewater and storm water overflow into the river. During low flows, all of the wastewater is collected and treated. This configuration was more feasible than installing new sewer lines from existing sources.

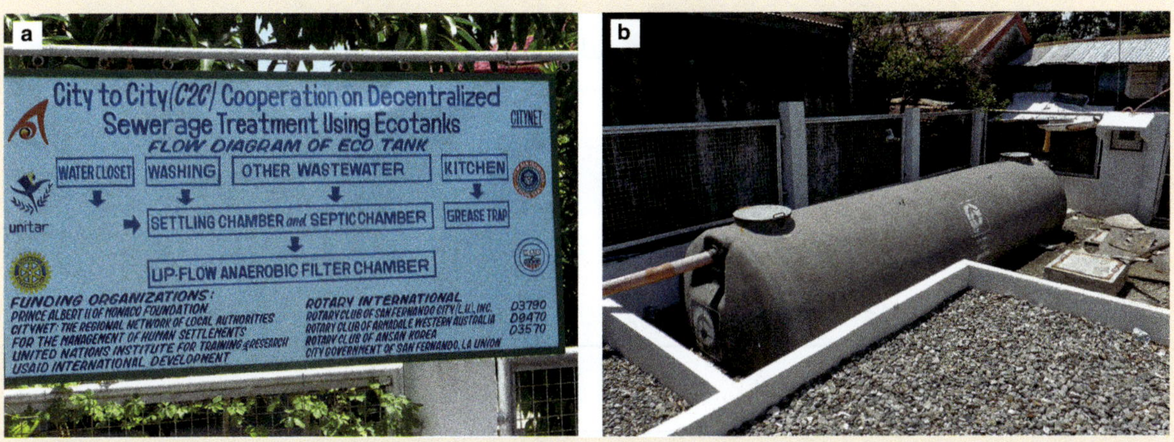

Figure 2.7 Community-based DWMS using EcoTank™ interceptor tank followed by upflow anaerobic filter, San Fernando City, Philippines (Authors).

This interceptor system uses a pump and capture basin to send the wastewater to the treatment tank (the EcoTank). During normal conditions, the entire combined wastewater and storm water flow is pumped to the EcoTank, a septic tank like-unit made by Premier Products Co. of Thailand. The EcoTank drains to a secondary treatment gravel filter. During storm conditions, the tank is bypassed and wastewater discharges directly to the river. Dilution of the sewage meets all regulatory standards during discharge events. This system is a good model for capturing and treating wastewater from un-sewered areas and shows the value of interceptor sewerage systems, which was recognized by city leaders, who are planning to use this model in other communities.

2.4 DETERMINING THE DESIGN FLOW

Determining the design flow is key to the DWMS planning process, as it influences the size of all components for sewer collection, treatment, dispersal, and reuse or discharge. The three methods for determining the design flow, described in the following sections, are as follows:

(1) measuring the flow of an existing facility by using actual water consumption data from water meters or other measuring devices such as buckets, and adjusting it through a wastewater conversion factor;
(2) comparing a proposed source's flow with known data from an existing similar facility; and
(3) using standard tables and values for wastewater generation for the unit processes occurring at the facility.

2.4.1 Measuring Flow

The first option for determining the design flow for existing facilities that are connected to metered water supply systems is relatively easy. For metered systems, the design flow is *the average daily water consumption (daily flow rate) during peak flow conditions for a consecutive 7-day period.* If outdoor piped water-using activities such as irrigation and car washing are common, this figure is then adjusted by applying a wastewater conversion factor (WCF) to account for water received but not converted to wastewater.

It is recommended that the peak 7-day period daily flow rate be obtained from a minimum of 3 months of data during peak usage. Given that design flows should be conservative, calculating for peak flows helps to ensure that wastewater is properly managed during times of the highest usage. Peak flow estimates may be made more precise by following the engineering convention of determining 1- and 4-hour peak loads. In general, the basic procedure for determining the average (7-day peak) daily water consumption requires taking daily meter readings for the three month period and writing them on a form. Including the concept of capacity (occupancy), and dividing the water consumption by that fraction as shown in Table 2.2, provides the adjusted water consumption data needed to derive the 7-day averages.

Table 2.2 Example procedure for determining peak 7-day water consumption from a hypothetical source.

Day	Water consumption (cubic meters)	Percent capacity of facility at the time of measurement (%)	Adjusted water consumption at 100 percent capacity	Running 7-day average
1	3.5	75	4.7 (from 3.5/0.75)	
2	4	80	5.0	
3	4	80	5.0	
4	4.2	85	4.9	
5	5	90	5.6	
6	5	90	5.6	
7	6	95	6.3	5.3
8	5.5	92	6.0	5.5

Guidelines for determining design flow from average peak 7-day water consumption data

- Always take the water meter readings at the same time each day.
- Before accepting the peak 7-day average water consumption figure flow rate (the highest value in the 'Running 7-day average" column) as the design flow, as a check to verify that it is appropriate, determine the average flow during the overall period tested (3-month minimum). The average flow should not exceed 70 percent of the peak 7-day average. If it does exceed that, adjust the preliminary peak value to meet the 70 percent condition by taking it to be the product of the average 3-month flow and (10/7). Finally, apply the appropriate WCF value as instructed in Section 2.4.1.1. below.

For facilities without a metered water service, the methodology for calculating the design flow is more difficult. It requires carefully looking at water-consuming activities in the facility, estimating the volume of water consumed during that activity, multiplying that by the number of times that activity is performed during a typical day in the peak month, and summing those values for all activities to get total daily flow. For example, if a pour flush toilet is to be connected to the DWMS, determining the volume of water used during each flush and multiplying by the number of flushes in a day (for all of the facilities' toilets) provides the daily flush volume, a major factor in most DWMS. While this method is less accurate than using water meter data, it is simple as it does not require using the Wastewater Conversion Factor, described below.

2.4.1.1 The Wastewater Conversion Factor

The WCF assesses how much of the total water entering a facility is actually converted to wastewater that flows or would flow to a DWMS. For facilities with piped water under pressure, this generally relates to the percentage of water use for activities that occur outdoors, such as landscape watering, garden irrigation, or vehicle washing. If these are frequent activities, the system planner should estimate the WCF in order to decrease the design flow to reflect how much of the initial metered flow does not enter the building sewer. This allows for the installation of smaller, more-appropriate and cost-effective installations.

The simple calculation for finding the WCF (for the peak 7-day period) is presented below. Once obtained, multiply it by the 7-day peak average flow to derive the design flow:

$$(\text{Total Weekly Metered Water } [m^3] - \text{Water Consumed for Outdoor Purposes } [m^3]) \times 100\%/\text{Total Weekly Metered Water } [m^3] = \text{WCF}.$$

In most situations, calculating a WCF is advisable, rather than using standard values, because WCFs are based on site-specific economic and climate conditions. If the site user does not use much piped water for watering the garden or washing the car – two major examples of outdoor use – most of the measured flow will be converted to wastewater, meaning the design flow is very close or equal to the measured flow. The same can be true even if the site has a demand for outdoor water use, because if the climate is cool and wet, then outdoor watering will be minimized.

This weather effect was found in a U.S. residential water-use study across 12 cities and over 1000 user accounts. For communities in hotter, drier climates, outdoor water consumption came to between 59 percent and 67 percent of total consumption, while that range was 22 percent to 38 percent for homes within cooler, wetter climates. Overall, the average percentage of metered water that went towards outdoor use in the study sample was 58 percent (Mayer *et al.,* 1999). This shows how the WCF – calculated here as 1 minus 0.58, which gives 0.42 – can be very high and greatly decrease design flows compared to the measured flows, putting much less of a burden on a proposed DWMS.

Calculating this important WCF term is simple and can be done quickly following the procedure shown in Table 2.3. The sample outdoor water consumption tracking table shows how data can be entered and multiplied to derive liters per activity values. Adding these up for the peak 7-day period sample used in Table 2.2 and dividing by 1000 to get cubic meters (the final volume unit) for weekly activities gives the 'Water Consumed for Outdoor Purposes' value, which is the only unknown in the WCF equation.

Table 2.3 Monthly activity log for tracking outdoor water consumption.

Date	Activity	Duration	Measured or estimated flow (L per minute)	Estimated consumption (L per activity)
2 March	Washing the car	30 minutes	10	300
5 March	Irrigating the landscaping	45 minutes	10	450
8 March	Irrigating the garden	25 minutes	10	250

2.4.2 Gathering and Applying Data from Similar Projects

A second option can be used if actual water consumption data are absent. This quick method is to estimate design flow by researching similar facilities to see if their water consumption is metered and can offer good estimates. Take care to find the proper comparison. Look for the numbers of family members or patrons served, the relative water demand, the flow changes over time, and other aspects of water use. If the comparison system was effectively planned, established, evaluated, and monitored, studying the system for its flow characteristics could add more value than just analyzing the water meter readings or using values from standard tables, which is described below.

2.4.3 Using Standard Wastewater Generation Value Tables

The third method of determining the design flow for DWMS is using standard tables that estimate wastewater generation per unit, such as the number of bedrooms in a residential house, number of employees in a shopping center, or number of occupants in a hotel. While actual values for wastewater generation can be significantly different from one location to the next, or even between similar facilities in the same location, standard tables may provide a strong basis for customizing values for local conditions. Table 2.4 provides U.S. standard wastewater generation values in liters per unit for typical residential and non-residential applications. These values may or may not reflect actual flows in developing country settings, but represent a good starting point.

Table 2.4 Wastewater generation by activity (adapted from Crites & Tchobanoglous, 1998).

Facility	Unit	Range (liters/day)	Typical (liters/day)
Airport	Passenger	8–15	11
Apartment house	Person	150–300	190
Auto service center	Vehicle served	30–57	45
Bar	Customer	4–19	11
Boarding house	Person	95–230	150
Department store	Toilet	1500–2300	1900
Hospital/medical	Bed	470–910	630
Hospital/medical	Employee	19–57	38
Hotel	Guest	150–230	190
Industrial building (sanitary waste only)	Employee	26–61	49
Laundry	Machine	1700–2500	2100
Office	Employee	26–61	49
Public toilet	User	11–23	19
Restaurant (per meal)	Meal served	8–15	11
School	Student	38–76	57
Shopping center	Employee	26–49	38
Theater	Seat	8–15	11

Using standard values to determine wastewater volume

- When using standard tables for wastewater generation, such as Table 2.4, note that the wastewater conversion factor is already built in, as are short term peaking factors. Simply multiply the number of units by flow per unit, and sum these values, to calculate the design flow.
- Local governments may have good sources of data on water consumption or wastewater generation. Discuss the project with local regulatory officials and determine if they have useful data that can be applied to the project.

2.5 ASSESSING WASTEWATER STRENGTH AND ORGANIC CONCENTRATION

Wastewater strength is based on the concentration of several parameters. While commercial or industrial wastewater sources may contain high levels of chemical contaminants or suspended solids, the typical concern of wastewater strength is organic loading. Although there are various definitions of strong and weak wastewater, which are all relative to the wastewater source and local regulation context, the values shown in Table 2.5 offer good baseline values.

The table shows values taken from numerous municipal wastewater studies in the United States. The table is useful for comparing wastewater strength from a project to these average values, and it may suggest problems or strategies for improved management. For example, having weaker than average wastewater may indicate excessive infiltration or the entry of excessive potable water into the wastewater (perhaps from leaky toilets). Excessively strong wastewater compared with average values may indicate a need to implement pretreatment technologies to reduce organic matter at the source.

Standards for relative strength of wastewater also can be used to identify appropriate technologies. For example, septic tanks are most applicable for low- to medium-strength wastewaters, while ABRs might be a better choice for medium- or high-strength flows.

Table 2.5 U.S. typical municipal wastewater characteristics, with values in mg/L (adapted from National Small Flows Clearinghouse, 1997).

Parameter	Weak	Medium	Strong
BOD_5	110	220	400
TSS	100	220	350
Nitrogen (N)	20	40	85
Phosphorus (P)	4	8	15

Determining organic loading is also important because it is often used in determining system sizing, and it can limit the effectiveness of many technologies, which are best used for wastewaters with certain BOD ranges. As an example, for sewage lagoons, BOD volumetric and surface area loading rates are used as key sizing parameters. Underestimating organic loading can result in mistaken pond sizing and total system failure. To avoid this, it is necessary to properly estimate organic loading levels. As with assessing design flow, this can be done either by preparing wastewater samples for BOD laboratory analysis, using standard values, or applying data obtained from operators of similar facilities.

For existing facilities, laboratory analysis for BOD_5 can be performed following the sampling and analysis guidelines provided in later sections. For new facilities, identifying a similar facility and sampling its wastewater for BOD can provide useful information. If both of these methods are not feasible, using standard values related to facility units (Table 2.6) can provide useful information. The values in the table can be used to estimate BOD concentration as follows:

$$BOD_5\left(\frac{mg}{L}\right) = \frac{\text{Table 2.6 value} * \text{number of units} * 1000\,\text{mg/g}}{\text{Flow}\left(\frac{L}{d}\right)}$$

Table 2.6 Typical BOD_5 loadings at various types of facilities (adapted from Onsite Sewage Treatment Program Staff, 2011).

Type of facility	BOD (g/unit/day)
Apartment block	79.5/housing unit
Boarding house	63.6/person
Church, no kitchen	9.1/seat
Country club	94.5/member
Dwelling, single family	77.3/person
Employee, additional	18.2/employee
Factory, no showers	33.2/employee
Factory, with showers	37.7/employee
Hospital	235.5/bed
Hotel	56.8/room
Office building, no food	22.7/employee
Restaurant	9.5/customer
Restaurant, with bar	14.1/customer
School, with cafeteria	19.1/student
Shopping center, no food	22.7/employee
Stores	378.2/toilet room
Theater	4.5/seat

Checking in with LORMA – Estimating design flow and organic loading

Because water meter and BOD_5 data were not available for estimating flow, standard values were used from Tables 2.4 and 2.6 to estimate organic loading at LORMA. LORMA is a modern hospital, so values from standard tables were used without adjustment, but were taken from the low end of the range for design flow. Design flow (in L/d) and organic loading were calculated as follows:

19 L/employee/d (Table 2.4) * 450 workers	= 8850 L/d;
470 L/bed/d (Table 2.4) * 136 beds	= 63,920 L/d;
Total Daily Design Flow	= ***72,770*** **L/d.**
18.2 g BOD_5/employee/d (Table 2.6) * 450 workers * 1000 mg/g/72,770 L/d	= 113 mg BOD_5/L;
235.5 g BOD_5/bed/d (Table 2.6) * 136 beds * 1000 mg/g/72,770 L/d	= 440 mg BOD_5/L;
Total Daily Organic Loading	= ***553*** **mg BOD_5/L.**

2.6 QUANTIFYING FLOW VARIABILITY

Flow variability is an important consideration in DWMS technology selection and a crucial factor in the source characterization process. Hotels, resorts, and schools are good examples of how wastewater flow volumes can change dramatically based on the day of the week, season, special events, or holidays. The variability can be identified by studying water meter readings if available, or through methods referenced above to determine design flow. This information will help designers choose appropriate technologies or strategies that may be used to equalize the flow to reduce peaks and enable designs based on average flow conditions.

2.6.1 Flow Equalization

Flow equalization refers to the usage of large tanks and effluent pumps to collect, store, and evenly discharge timed doses of wastewater to DWMS components at regular intervals. This is often beneficial because it:

- allows designers to base system designs on average rather than peak flows; and
- uniform flow conditions are easier to treat, resulting in better performance of treatment equipment.

Flow equalization example for a church

As with schools, churches or other places of worship may have distinct wastewater flow patterns during the week. Notice the difference in weekday versus weekend flows in the example below. All but the 3 days listed have flows close to zero. The flows can be equalized as follows:

Day	Attendance	Daily Flow
Sunday	200 members	4000 L
Wednesday	50 members	1000 L
Saturday	100 members	2000 L

Weekly total = 7000 L
Equalization cycle = 1 week (7 days)

Equalized daily flow = 7000 L/7 days = 1000 L/d

Assuming that this is the typical 7-day period for the church, the design flow equals 1000 L/d. Compare this with designing for the peak flow of 4000 L/d. Having the ability to design septic tanks and other DWMS components for the average flow, in this case 1000 liters per day, compared to the peak flow of 4000 L/d, requires a much smaller and cheaper DWMS, even when considering the cost of the storage tank and the pump.

Equalization is typically used for flows that have large variability throughout the day or week. The concepts can sometimes be applied on a much larger scale for flows with seasonal variability. For example, storage ponds can capture the entire wastewater flow from the peak season and release it evenly over the remaining low- or no-flow months. Depending upon the goal of the system, especially if the treated effluent is intended to be reused for irrigation, this scenario may be desirable as effluent can be stored for use during the growing season.

Flow equalization may be incorporated into a number of designs. However, estimating the degree of flow variability based on design flow data obtained earlier in the source evaluation process is first required. An example of how design flow and flow equalization are related is given below, while further technological considerations for flow equalization are provided in Chapter 4.

2.7 NUTRIENTS IN WASTEWATER

Nutrients are molecules that bacteria, plants, and animals need to live and grow. In wastewater, nutrients such as nitrogen (N) and phosphorus (P) can help support the growth of the beneficial microorganisms that accomplish the treatment process. However, high levels of certain nutrients in wastewater can pollute aquifers and surface waters, which can have significant health impacts on those who rely upon them for domestic needs.

Because nutrients are found in wastewater and are also required for plant growth, it is no surprise that treated wastewater is often used as an important source of nutrients for agricultural purposes. When safely applied, treated wastewater effluent can supplement or replace chemical fertilizers. In a world with rapidly increasing populations, the link between recycling nutrients from wastewater and achieving food security will likely become even stronger. Therefore, recycling the nutrient-rich treated effluent from wastewater is an important strategy for DWMS. Especially in agricultural settings, this could be a good reason or motivating factor that might encourage people to invest in the installation of a wastewater system.

2.7.1 Nitrogen

Urine is a major source of N in wastewater, which when discharged is almost completely in the form of ammonia. When ammonia mixes with atmospheric or soil oxygen, it can be converted into nitrate through a microbial process called nitrification. While some of the nitrates are naturally consumed through another microbial process called denitrification, nitrate accumulation in aquifers can occur when uncontrolled discharge of wastewater effluent enters the aquifer.

Nitrate contamination of aquifers that are used for drinking water can lead to serious health concerns. High nitrate concentrations in drinking water have been linked to certain forms of cancer, as well as to the disease known as methemoglobinemia (blue baby syndrome). For this reason, appropriate setbacks (minimum horizontal and vertical distances) between wastewater discharge points and wells and aquifers must be maintained.

Total Kjeldahl Nitrogen, which is the sum of ammonia-nitrogen plus organically bound nitrogen, is often associated with quaternary ammonia disinfectants used in large quantities in some commercial facilities. This occurrence presents another concern that can be a very important consideration in DWMS design.

2.7.2 Phosphorus

Phosphorus (P) is required for the life cycle of aquatic plants, such as algae, which can be either a health risk to aquatic life or simply unattractive to those living near or using waterways. In the case of blue-green algae, toxic byproducts can be generated. This creates human health issues if they are growing in water that is used for drinking, and harms species living in the water bodies.

P from detergents and fertilizers and N from fertilizers and human waste both present large challenges for wastewater managers worldwide. The widespread usage of N and P, and the negative effects on ecosystems from improper treatment or dispersal, make them a key consideration in the DWMS planning process. However, many governments in developing countries have yet to implement nutrient standards. Once they do, a large percentage of DWMS managers will have to quantify and reduce the high levels of these nutrients in their systems to meet regulatory standards and protect public health. However, determining the types and levels of nutrients in treated effluent can come in handy in and of itself, especially if agricultural reuse is planned.

More information about N and P in wastewater, and about the general processes for removing them can be found in Sasse (1998), pp. 50–52 and 62–63.

Planning DWMS for nutrient reduction

Designers of DWMS in clean environments, near important fisheries, or where groundwater is used for potable water must account for nutrient concentrations in the treated effluent when planning dispersal, discharge or reuse activities. Removing nutrients from the effluent through uptake in agricultural or landscape plants may be the best and most cost-effective control measure. Consider the following:

- Several DWMS technologies can reduce nutrient loads, and even convert some of them into plant biomass, as described in Chapter 4.
- Land application of properly treated wastewater effluent must be carefully monitored to verify that the proper amount is being applied, and that it does not run off, which could easily contaminate surface waters.

Case Study – Nutrient recycling at Putatan Elementary School, Muntinlupa City, The Philippines

Wastewater from Putatan Elementary School in The Philippines is first treated in a cocopeat bioreactor system. Once treated, the effluent is discharged into the garden to irrigate a variety of food and non-food crops (Figure 2.8). The students and staff noticed immediately that the plants that received the treated effluent grew much faster and were much healthier than those that received potable water. According to J.E.T. Pabilonia, principal researcher on the cocopeat project, 'Recycling the nutrients in wastewater effluent by irrigating plants in moderate levels is one way to achieve two goals: keeping the nutrient pollution out of the environment, and using it to produce valuable crops'.

Figure 2.8 J.E.T. Pabilonia at the Putatan Elementary School wastewater demonstration project, Muntinlupa, Philippines (Authors).

2.8 OTHER WASTEWATER POLLUTANTS

Three additional wastewater pollutants must be considered when designing DWMS for commercial facilities. These are fats, oils and grease from food service or food processing facilities; lint from commercial laundries; and trash, which may include solid waste entering the waste stream, or vegetable-preparation, meat trimming, or fish scaling waste typically generated at public markets. These topics are further described below and in the section on pre-treatment technologies in Chapter 4.

2.8.1 Fats, Oils, and Grease

Fats, oil and grease (FOG) are common pollutants in wastewater. If left unchecked, FOG deposits can lead to clogging and fouling of downstream sewers and wastewater management equipment. Organized community-based programs that encourage FOG separation and collection can collect and provide commercial quantities of this important feedstock used for manufacturing goods including tallow and biodiesel. Also, the collected FOG can be used directly as a fuel source for heating or cooking. To manage FOG, grease traps should be used in any business that prepares or manufactures food, or where utensil washing occurs, such as restaurants, hospitals, resorts, and schools.

2.8.2 Lint

Lint is composed of particles of textile fibers primarily from cotton, wool, and linen. During clothes-washing, lint builds up in the wash water and can clog downstream drainage pipes and wastewater treatment equipment. In commercial laundry operations, lint accumulation can make up a significant volume of the sludge in septic tanks. Removal of lint at the source through lint traps is a best management practice. Use lint traps in any business that conducts laundry washing including public laundry facilities, hospitals, hotels, and boarding schools.

2.8.3 Trash

In the context of wastewater treatment, trash is a broad category of solid waste that may end up in the sewer system, either through toilets or open sewer channels. Businesses or other institutions that generate solid debris should use solids separators to trap the materials before they can enter the sewer line. Public markets are good examples where operations related to vegetable, and meat and fish sales can generate high volumes of solids in the wastewater. Specially designed in-line solids or trash traps should be used for such facilities.

2.9 PHYSICAL AND CHEMICAL CONSIDERATIONS

Commercial wastewater dischargers may be required to pretreat wastewater before discharge into the collection system for a DWMS due to extreme conditions related to physical or chemical properties of their wastewater. Proper pretreatment to address such properties (and the Section 2.8 constituents) is necessary because it:

- protects the environment, public health, and safety;
- provides safe working conditions for sewer utility workers;
- protects wastewater, collection, and treatment facilities from damage due to accidental or deliberate discharge of pollutants;
- prevents the introduction of substances that would block or obstruct the collection system;
- prevents the introduction of pollutants into the collection system, which would interfere with the wastewater treatment process or pass through the plant and pose a health threat; and
- improves the opportunity to reclaim and recycle municipal wastewater and biosolids.

The following sections provide related guidance for commercial and industrial dischargers to DWMS that relates to what are often the most problematic physical and chemical conditions in certain wastewaters: temperature, pH and the presence of industry-specific chemicals.

2.9.1 Temperature

The temperature of wastewater influent is a major factor in the rate of DWMS biological function. At higher ambient temperatures, microbial activity accelerates, which means a greater ability to break down pollutants. This improves the functioning of biological DWMS filtration methods. Cold temperatures decrease microbial activity and lengthen the amount of time that wastewater must undergo biological processing. Designers of wastewater systems based in cold or temperate climates must take this factor into consideration. Sewage lagoons and anaerobic digesters are good examples. A lagoon system that might require 45 days to achieve proper treatment in a tropical climate might require 60–90 days in temperate or cold climates. Follow design guidance for temperature variability, which is discussed in greater detail in Chapter 4.

2.9.2 The pH Level

The acidity or basicity of a liquid is defined as its pH, which ranges from the most acidic value of zero to the most basic value of 14. Seven, the pH of water, is considered neutral. Understanding how pH affects biological wastewater treatment processes, specifically the bacteria that degrade organic matter, solids, and pathogens, is important to the success of those systems. Those bacteria best operate at a pH around 7. However, wastewater from certain commercial and industrial sources may have discharges with extreme pH values, which can slow down microbial activity and greatly decrease treatment performance. In some cases, having a pH greater than 7 is actually advantageous to the treatment process. One example is where pH values of around 8.0 are optimal for nitrification to occur, which is often carefully planned for wastewater management where N is a major concern. However, in general pretreatment strategies for acidic or basic wastewater center on neutralization, which is the addition of enough acids or bases to drive the pH of the wastewater to as close to 7.0 as possible. Lime is often used as a buffering agent to help acidic solutions reach more neutral pH values.

The pH level is tested by dipping a handheld pH meter or test strip into the water and observing the value from the meter or the color change on the test strip that is then compared to a pH scale. For residential-type wastewater, typical monitoring includes pH testing, which might help to identify if household chemicals are being used excessively and discharged down the drain. For commercial flows, careful monitoring of pH can help spot problems such as microbial die-off before they become very serious and lead to poor DWMS performance.

2.9.3 Problematic Chemical Constituents in Wastewater

Wastewater system planners and operators must also be careful of chemical discharges containing elements besides N and P. Products such as bleach, livestock-injected growth hormones, and other commercial or agricultural substances can also harm or kill the microbes that drive biological wastewater processes occurring within DWMS technologies such as biofilters and biodigesters.

Chemicals from commercial laundry operations, metal plating, and leather tanning pose major threats to DWMS function. Dry-cleaning facilities may use chemicals that can be categorized as acids, bases, oxidizers, solvents, or other organics. All of these can be toxic to microbes. In some cases, they should not mix with one another, as mixing could cause adverse reactions or even explosions, such as when chlorine and ammonia combine. Proper operation of these facilities requires completely separating these chemical wastes from the rest of the wastewater stream.

The most basic chemical contamination of biological wastewater treatment is due to the antibacterial effects of chlorine, the major component of bleach. Anaerobic systems such as septic tanks and anaerobic baffled reactors often need to neutralize typical amounts of household bleach used in residential settings. However, facilities that use larger amounts of disinfectants, such as hospitals, health care centers, and even slaughterhouses, may need to use appropriate pretreatment devices to protect downstream treatment systems.

2.10 MICROBIOLOGICAL CONSIDERATIONS

There are many types of microorganisms in wastewater. Some are responsible for the breakdown of organic matter in the wastewater and are considered beneficial to the wastewater treatment process. Others are pathogenic or potentially pathogenic, and can pose significant risks to public and environmental health in developing countries. This suggests the need to evaluate the concentrations of pathogenic microorganisms in the raw wastewater as well as the allowable concentrations based on discharge and effluent reuse activities (examples shown in Section 4.9.1) to ensure that appropriate treatment technologies are utilized.

Pathogenic microorganisms may include viruses, bacteria, protozoan parasites, helminths, fungi, and algae, most of which are shed in human feces. As it would be difficult and costly to quantify each pathogen in a wastewater sample, investigators use indicator organisms to determine if water or wastewater contains human sewage and the concentration of potentially pathogenic organisms. Also, many developing countries have adopted standards for the discharge or reuse of treated effluent, which are also based on these indicator organisms.

In addition to helminths and rotavirus, which are used within the World Health Organization's 2006 Guidelines for the Safe Use of Wastewater, Excreta and Graywater, as described in Section 4.9.1, one or more of the three bacterial indicators listed below are often used for setting wastewater effluent or reuse standards:

- *Total coliform bacteria*: a broad group of bacteria that are not typically dangerous to human health, but are not naturally present in groundwater. Presence of total coliform bacteria is an indication that more harmful organisms may be present;
- *Fecal coliform bacteria*: a subset of total coliforms as shown in Figure 2.9 that can indicate recent fecal contamination; and
- *Escherichia coli bacteria*: a subset of fecal coliforms as shown in Figure 2.9 that usually indicates recent fecal contamination.

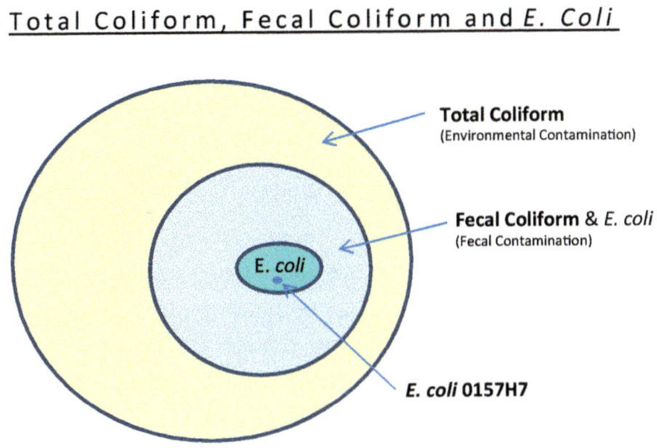

Figure 2.9 Model showing relationships between coliform indicators (adapted from Washington State Department of Health, 2013).

Compartment Bag Test kits for *E. coli* sampling – innovative rapid and low cost microbiological testing for drinking water and low-strength wastewater

The Aquagenx Compartment Bag Test (CBT) (Figure 2.10) is an easy to use and affordable test kit for determining the concentration of *E. coli* in water. It is hand-held, safe, and requires no laboratory or expensive incubator equipment – needed to raise the liquid's temperature to maximize pathogen replication for ease of counting – in temperate or tropical conditions. It just requires ambient temperature incubation, after which the number and order of compartments (filled in with varying volumes of water) that turn blue are 'read' to determine Most Probable Number of *E. coli*.

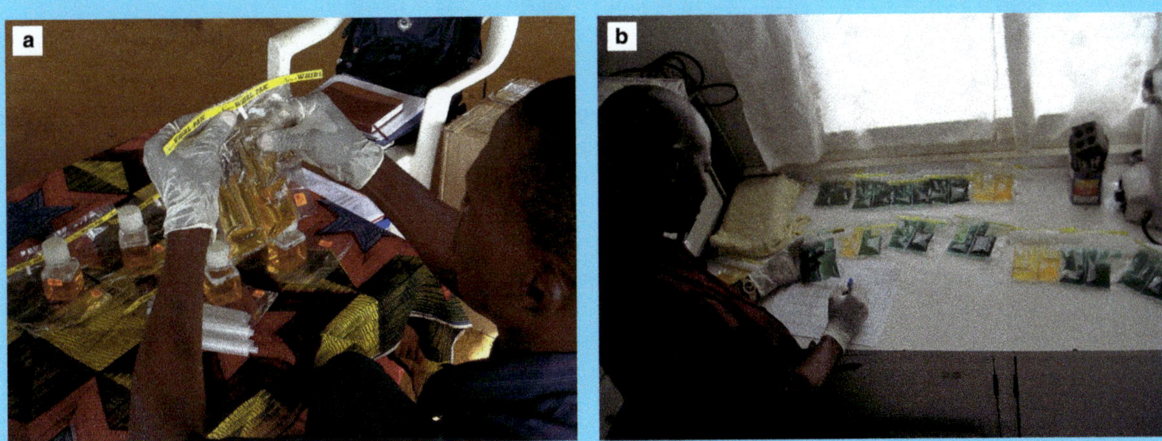

Figure 2.10 Worker in Liberia first prepares samples for *E. coli* analysis using the CBT, and then scores results to obtain the most probable number (MPN) of *E. coli* bacteria (Aquagenx, 2013).

A United States Agency for International Development pilot study with the CBT across diverse geographical areas (coast, jungle, and mountains) determined that CBT results are not statistically different when compared to other standard methods of *E. coli* detection and quantification. The test can be used at the household level, for water utilities to determine the effectiveness of their disinfection programs, and for testing reclaimed water intended for agricultural purposes or weak wastewater (by using a series of dilutions).

For more information, visit Aquagenx online at: http://www.aquagenx.com/

Although it is certainly easier to sample only these fecal indicators rather than test for every potential pathogen of public and environmental health importance, this is not to say that such sampling is simple. The following discussion will describe the basics of the general process, while the below case study offers one example of an increasingly-used simple device designed for sampling waters in low-resource settings.

2.11 SAMPLING AND ANALYSIS

The sections above present several methods for assessing wastewater flow, strength, and concentration of common pollutants typically found in wastewater. It is often useful, especially in larger systems, to collect actual data on wastewater flow, strength and pollutant loading to have a more accurate basis for making design decisions. This can be done through sampling and laboratory analysis. While this may sound simple, several pitfalls must be avoided to obtain accurate and usable data. This section describes sampling and analysis strategies that DWMS designers should consider using to accurately characterize their wastewater sources.

Given that field sampling and laboratory analysis are expensive, care must be taken to ensure that the funds are spent wisely and that the generated data is not only accurate, but useful. To maximize the effectiveness of the data collection activities, the following steps will be described:

(1) preparing a sampling plan;
(2) collecting representative samples;
(3) handling and preserving samples properly;
(4) using appropriate chain-of-custody and sample identification procedures; and
(5) assuring quality of the procedures in the field.

2.11.1 Preparing a Sampling Plan

The sampling plan is a written document that describes:

- why the samples are required;
- what information will be obtained;
- who will collect the samples in the field, and how and when he or she will do so;
- where the samples will be taken to for analysis; and
- how samples will be safely preserved and transported to the laboratory.

The purpose of the sampling plan, which should be prepared regardless of the size of the sampling project, is to ensure that all aspects of the sampling events are properly considered. This achieves the ultimate goal, which is to *safely collect representative samples of the wastewater to be monitored, properly preserve the samples, and get them to the laboratory within the required hold times*.

Coordinating sampling with the selected laboratory

The laboratory that will be responsible for conducting the analysis should be involved at the earliest stages of the planning for the sampling event. The laboratory should provide the bottles, preservatives, labels, and forms. Often, the laboratory can provide trained sampling personnel for a small fee. For the best results, use a laboratory licensed by a local or national regulatory agency. Verify with the laboratory each aspect of the sampling plan to assure conformance prior to the sampling event.

2.11.2 Collecting Representative Samples

For all but the most complex sampling programs, wastewater samples may be collected by the following methods, listed in ascending order for accuracy and cost:

- *grab sample* – a single sample taken at a specific time;
- *composite sample* – a single sample composed of two or more portions representing a waste stream over a period of time; and

- *flow-weighted composite sample* – a single sample composed of two or more portions where the portion size is proportional to the flow at the time the portion was collected. As an example, if the flow at time A is two times that at time B, the volume of the sample portion for time A should be twice that of time B.

Grab samples should be used when wastewater does not flow continuously or when wastewater quality does not vary greatly. Take multiple grab samples at different times of the day and different days of the week to try to determine average contaminant concentrations relative to the variability of the flow. In contrast, use composite and flow-weighted composite samples when more accurate data are required or when wastewater constituent concentrations are highly variable. Composite sampling usually requires automated sampling equipment and flow measuring devices. Hire laboratory or trained personnel to conduct this sampling.

2.11.3 Proper Handling and Preservation of Field Samples

Wastewater contains infectious microorganisms, which means field samples must be collected and handled in a hygienic manner. Personal protective equipment including latex gloves (and, ideally, eye protection) must be worn at all times during sample collection and handling (Figure 2.11).

Figure 2.11 Technician preserves field samples with sulfuric acid (Authors).

Wastewater samples degrade quickly so they must be preserved properly for transit to the laboratory for processing. This must be accomplished within the hold time, which is specific for each test. The contract laboratory should provide instructions and materials for proper field preservation of samples. Check with the laboratory for specific details about these two primary methods:

- preservation in an ice chest at 4°C, good for pathogen/BOD sampling (hold time of 6 hours) and for total/volatile solids and other parameter testing, usually with 7-day hold times; and
- preservation through the use of certain chemical compounds, usually sulfuric or nitric acid, for N-, P-, and other chemical-related tests, most often with 7-day hold times (Fulhage *et al.*, 1993).

For more information about sampling and preservation methodology, refer to Fulhage *et al.* (1993) at http://extension.missouri.edu/p/G1895.

2.11.4 Chain of Custody and Sample Identification Procedures

Chain of custody tracks the movement of samples and who controls them from the time of sampling, through transportation to and receipt of the samples at the laboratory, and ending at the analysis stage. Laboratories provide samplers with Chain of Custody forms (Figure 2.12), which are legal documents that provide the following:

- sample identification;
- data related to the site conditions at the time of sampling;
- instructions to the laboratory about the requested analytical testing and storage of each sample; and
- the date, time, and signature of each person that maintains custody of the sample.

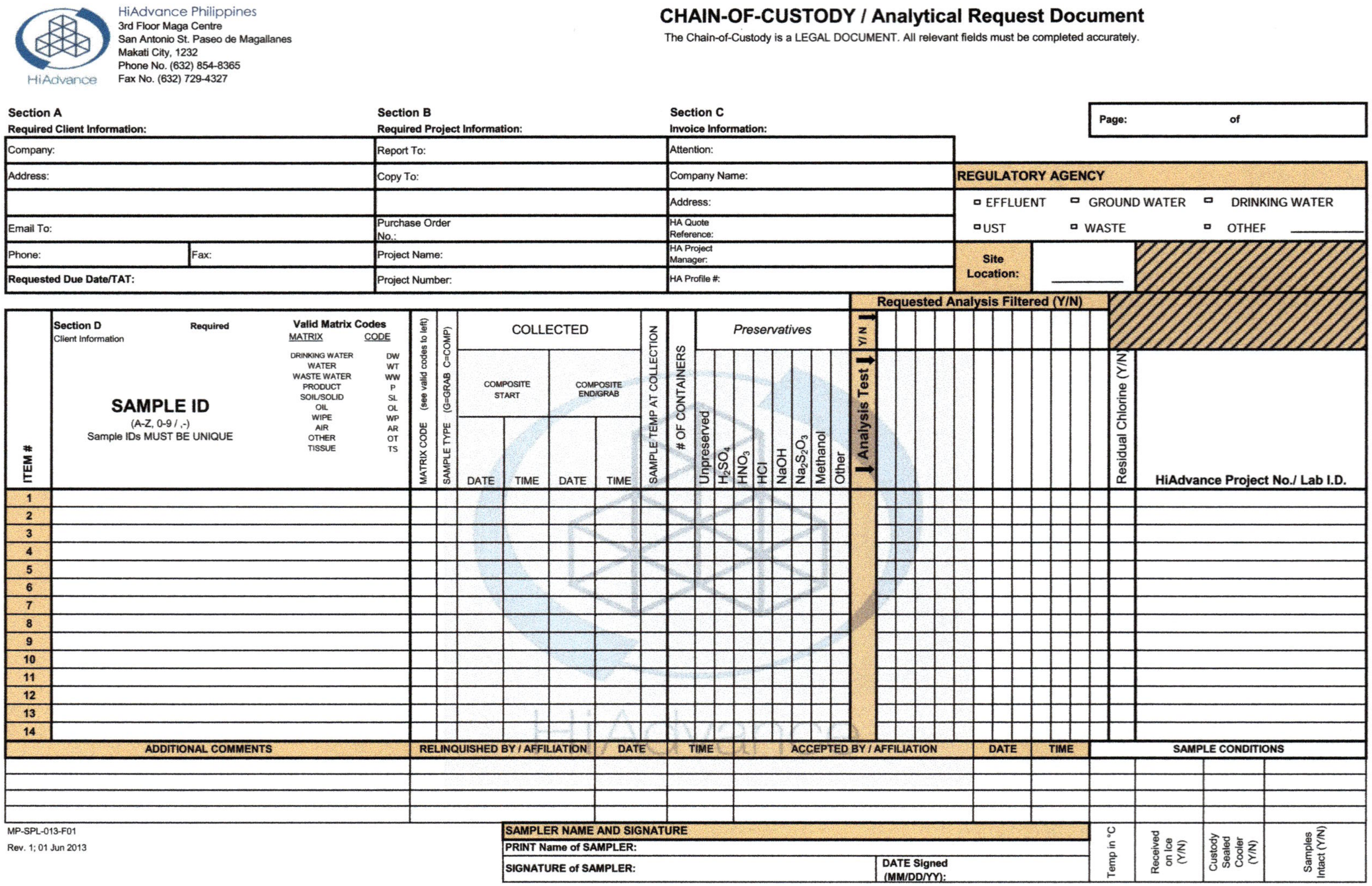

HiAdvance Philippines
3rd Floor Maga Centre
San Antonio St. Paseo de Magallanes
Makati City, 1232
Phone No. (632) 854-8365
Fax No. (632) 729-4327

CHAIN-OF-CUSTODY / Analytical Request Document

The Chain-of-Custody is a LEGAL DOCUMENT. All relevant fields must be completed accurately.

Page: of

Section A
Required Client Information:
Company:
Address:
Email To:
Phone: Fax:
Requested Due Date/TAT:

Section B
Required Project Information:
Report To:
Copy To:
Purchase Order No.:
Project Name:
Project Number:

Section C
Invoice Information:
Attention:
Company Name:
Address:
HA Quote Reference:
HA Project Manager:
HA Profile #:

REGULATORY AGENCY
□ EFFLUENT □ GROUND WATER □ DRINKING WATER
□ UST □ WASTE □ OTHER ________

Site Location: ________

Requested Analysis Filtered (Y/N)

Section D Client Information — Required

SAMPLE ID
(A-Z, 0-9 / ,-)
Sample IDs MUST BE UNIQUE

Valid Matrix Codes MATRIX	CODE
DRINKING WATER	DW
WATER	WT
WASTE WATER	WW
PRODUCT	P
SOIL/SOLID	SL
OIL	OL
WIPE	WP
AIR	AR
OTHER	OT
TISSUE	TS

ITEM #	SAMPLE ID	MATRIX CODE (see valid codes to left)	SAMPLE TYPE (G=GRAB C=COMP)	COLLECTED – COMPOSITE START DATE	COMPOSITE START TIME	COMPOSITE END/GRAB DATE	COMPOSITE END/GRAB TIME	SAMPLE TEMP AT COLLECTION	# OF CONTAINERS	Preservatives: Unpreserved	H_2SO_4	HNO_3	HCl	NaOH	$Na_2S_2O_3$	Methanol	Other	Analysis Test ↓ Y/N												Residual Chlorine (Y/N)	HiAdvance Project No./ Lab I.D.
1																														
2																														
3																														
4																														
5																														
6																														
7																														
8																														
9																														
10																														
11																														
12																														
13																														
14																														

ADDITIONAL COMMENTS	RELINQUISHED BY / AFFILIATION	DATE	TIME	ACCEPTED BY / AFFILIATION	DATE	TIME	SAMPLE CONDITIONS			
							Temp in °C	Received on Ice (Y/N)	Custody Sealed Cooler (Y/N)	Samples Intact (Y/N)

SAMPLER NAME AND SIGNATURE
PRINT Name of SAMPLER:
SIGNATURE of SAMPLER: **DATE Signed (MM/DD/YY):**

MP-SPL-013-F01
Rev. 1; 01 Jun 2013

Figure 2.12 Typical Chain of Custody form (HiAdvance Laboratories, 2013).

2.11.5 Quality Assurance/Quality Control

Quality assurance and quality control (QA/QC) refer to the policies, procedures, and activities established to ensure confidence in the accuracy of the data. QA/QC is applicable to field and laboratory procedures alike for wastewater sampling, as described in the following sections.

2.11.5.1 QA/QC in the Field

The following practices are recommended to minimize errors and ensure quality of field-based sampling events. As these are generally under the control of the sampler, they are presented here as best practices:

- use of standard operating procedures for sample collection and analysis;
- use of chain-of-custody and sample-identification procedures;
- instrument standardization, calibration, and verification;
- sampling technician and analyst training;
- assurance of appropriate preservation, handling, and decontamination; and
- use of QC samples such as field and trip blanks, duplicates, and equipment rinses.

2.11.5.2 QA/QC in the Laboratory

All laboratories should have QA/QC plans that should be up-to-date and available for viewing upon request. The laboratory's QA/QC plan should state the procedures that the laboratory uses to ensure that their internal practices are valid and follow industry standards. Customers of contract laboratories can request copies of a laboratory's QA/QC plans and verify that they meet minimum standards.

Use the sampling and analysis checklist (Table 2.7) to track key activities related to field and laboratory data collection.

Table 2.7 Sampling and analysis checklist.

Sampling Plan Check List	Y	N
There is a written sampling plan.		
The number of samples, the analytical tests, and the sampling locations have been identified in the plan.		
The laboratory that will do the analysis has been contacted, procedures have been reviewed, and services have been scheduled.		
All sampling bottles, sample preservatives, labels, ice chests, and Chain of Custody forms have been received.		
The person(s) who will conduct the sampling has been trained in proper procedures.		
Health and safety training, as well as the required personal protective equipment, has been provided.		
Access to the sampling locations is open and unrestricted.		
Transport of the samples to the laboratory has been arranged and will be done within the required hold times.		
The laboratory is licensed to conduct the required work and that they have a QA/QC plan.		

2.12 REVIEWING FINDINGS

Once the data collection and/or testing is completed, or the assumptions are made about the nature of the specific flow, review the findings. The wastewater should be described based on the following characteristics:

- flow rate;
- flow strength;

- flow variability;
- nutrients and other problematic wastewater pollutants like FOG, lint, and trash;
- physical and chemical considerations; and
- other source-specific conditions that might impact upon DWMS design or function.

After reviewing each characteristic, ensure that the results are documented for current and future purposes, and keep a written record of baseline conditions. These results provide a basis from which to evaluate changes, which may result from increased flow rates or changing conditions or processes in commercial and industrial facilities. Written records are also useful for compliance purposes.

Table 2.8 can be used to summarize the assumptions and findings. In this instance, the table is filled out with the data collected for LORMA and represents the basis of the DWMS design for their wastewater facility.

Checking in with LORMA – A review of source characteristics

Because the design flow and organic loading were already determined, a review of the other characteristics of the wastewater source was needed:

- *Variability*: There is minimal flow variability. Most of the beds are usually full and the facility is open 7 days a week, 365 days per year;
- *FOG*: There is food preparation, food service, and utensil washing at the hospital, so it can be assumed that FOG is generated. Therefore, some type of grease trap or interceptor will be required;
- *Lint*: There is laundry washing at the hospital, so lint management will be needed;
- *Trash, and hazardous and medical waste*: As a modern and major medical facility, it generates hazardous and medical waste. For LORMA, all hazardous and medical waste are collected, stored, and managed separate from the wastewater flow. They have closed sewers, so trash entering the wastewater stream is not considered a problem;
- *Nutrients*: There are no activities involving fertilizers or other N- or P- compounds, so no special focus is needed;
- *Physical and chemical aspects*: pH and especially temperature are not at extreme values, but bleach and chlorine surface-cleaning solutions are commonly used, which should be accounted for during primary and secondary treatment technology selection.

These conditions are listed in Table 2.8 and represent the findings of the source characterization study.

Table 2.8 Source characterization review and summary form (example from LORMA case study).

Parameter	Description	Value from source
Hydraulic loading	What is the wastewater flow rate (L/d)?	72,770 L/d
Organic loading	What is the BOD loading (mg/L) from the wastewater source?	553 mg BOD_5/L
Flow variability	Does the source flow vary from day to day, week to week, or seasonally, and suggest the need for flow equalization?	No
Fats, oil, and grease (FOG)	Are there food service, food manufacturing, or utensil-washing activities?	Yes
Lint	Are there commercial clothes-washing activities?	Yes
Trash	Are there special activities that introduce unmanageable solid trash particles or hazardous-related waste?	No – hazardous and medical waste are segregated
Nutrients	Are there any regulatory requirements for nutrient reduction or obvious concerns in general?	No
Physical aspects	Are there any unusual temperature or pH aspects of the flow due to commercial activities?	No
Chemical aspects	Are there any other chemical constituents from commercial activities that may be in the wastewater?	High levels of quaternary ammonia disinfectant are used

REFERENCES

Aquagenx staff (2013). Simple, Portable Water Quality Test. http://www.aquagenx.com/ (accessed 2 December, 2013).

Crites R. and Tchobanoglous G. (1998). Small and Decentralized Wastewater Management Systems. McGraw-Hill, Boston, Massachusetts, USA.

Fulhage J., Porter J. and Sievers D. (1993). Collecting and Preserving Waste and Wastewater Samples for Analysis. http://extension.missouri.edu/p/G1895 (accessed 10 November 2012).

HiAdvance Laboratories (2013). HiAdvance Philippines Incorporated, 3rd floor Maga Center, San Antonio St. Paseo de Magallanes, 1232, Makati City, Philippines.

Lindstrom C. (2000). Treatment Technologies. http://www.greywater.com/treatment.htm (accessed 22 November 2013).

Mayer P. W., DeOreo W., Opitz E., Kiefer J., Davis W., Dziegielewski B. and Nelson J. (1999). Residential End Uses of Water: A Study Funded by the American Water Works Association Research Foundation. http://www.waterrf.org/PublicReportLibrary/RFR90781_1999_241A.pdf (accessed 2 March 2013).

National Small Flows Clearinghouse (1997). Basic Wastewater Characteristics: Pipeline Newsletter. http://www.nesc.wvu.edu/pdf/WW/publications/pipline/PL_FA97.pdf (accessed 14 April 2013).

Onsite Sewage Treatment Program Staff, University of Minnesota (2011). Manual for Septic System Professionals in Minnesota, Chapter 5: Wastewater Sources and Flow. http://www.septic.umn.edu/sstsmanual/index.htm (accessed 15 February 2013).

Santos C. and Robbins D. (2011). Low-cost Innovative Solutions for Treating Public Market Wastewater in the Philippines: Deploying Hybrid Anaerobic and Aerobic Cocopeat Filtration Systems. http://www.rti.org/publications/abstract.cfm?pubid=18248 (accessed 12 April 2013).

Sasse L. (1998). DEWATS: Decentralized Wastewater Treatment in Developing Countries. http://www.sswm.info/sites/default/files/reference_attachments/SASSE%201998%20DEWATS%20Decentralised%20Wastewater%20Treatment%20in%20Developing%20Countries_0.pdf (accessed 5 October 2012).

University of Wisconsin-Madison (1978). Management of Small Wastewater flows (EPA 600/7-78-173). Office of Research and Development, Municipal Environmental Research Laboratory, US EPA, Cincinnati, OH.

Washington State Department of Health (2013). Coliform Bacteria in Drinking Water. http://www.doh.wa.gov/CommunityandEnvironment/DrinkingWater/Contaminants/Coliform.aspx (accessed 3 June 2013).

Chapter 3

Evaluating the site

This chapter will

- Introduce the site plan drawing;
- Present the step-by-step procedures for evaluating soils, which helps determine a site's potential for dispersing wastewater effluent;
- Discuss how to identify constraints related to groundwater and surface water;
- Describe the other factors studied within site evaluation, including land use of surrounding parcels, utility conflicts, and access; and
- Consider how the desire to reuse or recycle the residuals of wastewater treatment can drive technology decisions.

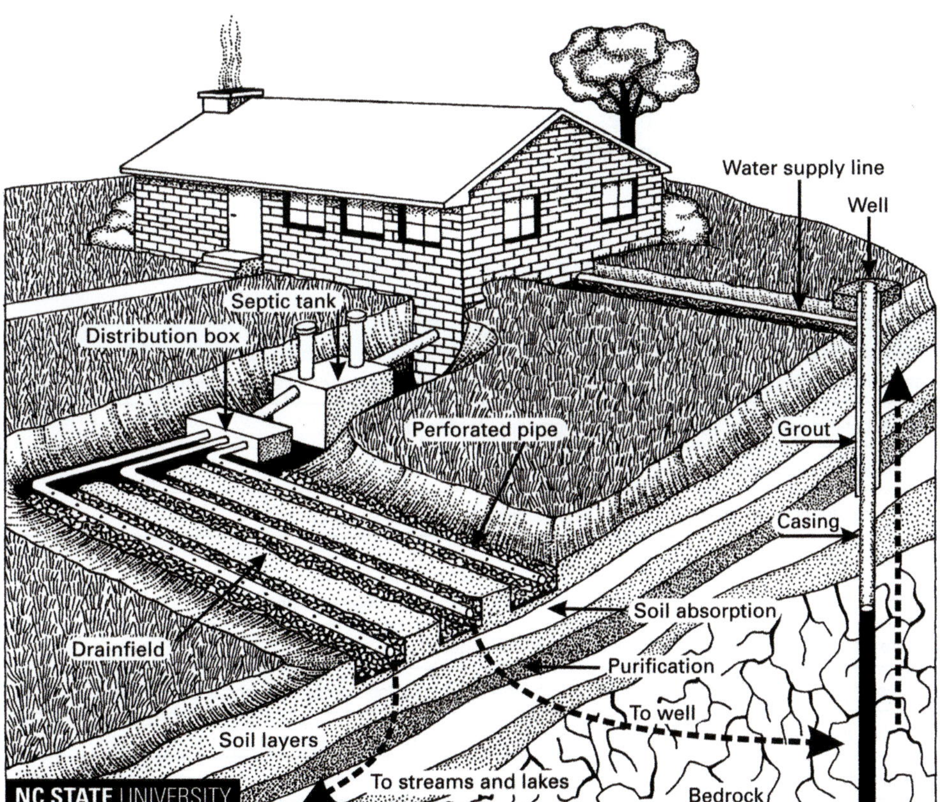

Figure 3.1 Crucial processes, site features, and concepts of soil-water passage that guide soils-dispersal DWMS technology selection as part of the site evaluation process (Hoover, 2013).

About Figure 3.1

Figure 3.1 depicts a single family home and how wastewater flows from the building, through the septic tank, and into a soil dispersal system. In this case, the soil dispersal system is a leach trench, which is a gravel-filled excavation that is sited and sized so that the effluent will receive proper treatment as it passes through the soils.

The image reminds us that what is above the ground is just a part of what must be considered when designing a watewater management system. Through the aid of soil discovery holes (test holes), we can uncover the different layers of soils that hold clues as to which management technologies might be best for a particular project. This chapter covers the evaluation of the surface and subsurface features of the site, which are used to make informed DWMS technology selections and to provide information (such as presence of groundwater or bedrock) to the installer before excavation occurs.

In many parts of Asia and Africa, septic tanks are installed without bottoms. Or, if two compartment tanks are used, the second compartment often has an open bottom to encourage leaching. While these leaching septic tanks may be effective at removing the wastewater from public contact, they may do little to treat the wastewater as it moves through the soil and into the aquifer. Indeed, in many parts of the world, such systems are directly responsible for contamination of aquifers that supply large portions of the population with drinking water. At a minimum, the materials provided in this chapter may help to dispel the idea that leaching septic tanks are sufficient forms of wastewater management and to provide ways to improve upon this choice.

The soils and other site evaluation techniques presented in this chapter do not require any specialized equipment. They can all be performed on site by trained practitioners. Once mastered, these techniques will enable the investigator to rapidly determine the adequacy of the soils for on-site dispersal and soils-based treatment and, if adequate, how much wastewater can safely be applied to a given land area. This knowledge can be used for a DWMS that either leaches the effluent away, or reuses it for landscape or crop irrigation. Understanding the soils is a key component of the site evaluation process and is a main feature of this chapter. Teaching this step-by-step methodology is one of the tools that can be used to build the capacity of local service providers to enable the widespread application of DWMS technologies in developing countries.

3.1 INTRODUCTION

The second set of necessary inputs in the DWMS decision-making process is obtained through evaluating the site where the wastewater management will take place. Preparing a site plan drawing and understanding key site features, such as soil properties, slope and topography, water resources, land use, nearby utility connection points, and overall site access are vital to choosing the most appropriate technologies. The sections and checklist Table 3.1 below can aid the site investigator in organizing activities and conducting them in a step-by-step process.

Table 3.1 Site evaluation checklist.

Task	Y	N
Prepare a scaled site plan sketch showing property lines, buildings, slope, and other features that can be used to estimate land area available.		
Excavate test holes or auger borings and assess soil horizon arrangement, depth and thickness.		
Evaluate the soils for structure, texture, consistence, and other properties.		
Determine the depth to groundwater or other limiting conditions and identify the shallowest such condition.		
Determine site slope and landscape position.		
Determine long-term acceptance rate (LTAR) for each soil horizon and identify other soils-based concerns.		
Determine surface water elevation and identify flood plains and surface drainages.		
Assess land use and key water features of the parcel and surrounding areas, including location of wells and surface waters.		
Locate utility connection points and any pre-existing DWMS structures on the property.		
Verify that the site is determined to be free and open, with any restrictions noted.		
Identify desired end uses of the treated effluent and residuals.		
Identify any regulatory compliance restrictions or zoning, land use, or discharge restrictions.		

3.2 CREATING AND USING A SITE PLAN SKETCH

A site plan sketch helps the service provider plan the key technology features of the DWMS. The sketch serves as a tool for describing all of the physical features of the site on one page. It is used to determine the locations of the site's existing and future structures and features, to estimate the available surface area for DWMS components, and to verify compliance with setbacks from wells and other features. It is also used to map the flow of raw wastewater as it moves from the source to the treatment system, through the treatment components, and finally to the point of discharge, reuse, or dispersal.

A site plan drawing (example for LORMA in Figure 3.2) identifies property lines, wells, setbacks, physical structures, buildings, slope and topographical information, floodplains and drainages, where soils tests were conducted, and locations on the site where the DWMS infrastructure can be installed, as well as gives slope and topographical information. It is drawn to scale and is fully dimensioned. Good site plan drawings are effective for communicating information about a project to regulators, contractors, stakeholders, or other interested parties.

In this LORMA Medical Center site drawing (provided for illustrative purposes only), notice the following features that correspond with the circled numbers on the drawing:

(1) Topography contour lines indicating the slope. In this instance, they are based on a fixed point which is assigned an arbitrary value of 100 M from which all slopes on the property are measured;

(2) Property dimensions. These can often be obtained from land use records at the local government, but when in doubt, a formal survey can help identify actual property lines;

(3) Test hole (soil boring locations). These are plotted on the site plan and help identify the variability of subsurface features;

(4) Locations of all wells and water lines on the subject property, and on neighboring parcels. Their locations are shows to verify compliance with setback requirements;

(5) Location of existing septic tanks. This helps them to identify the existing wastewater discharge locations, enabling the designer to effectively plan sewer lines and locations for other system components.

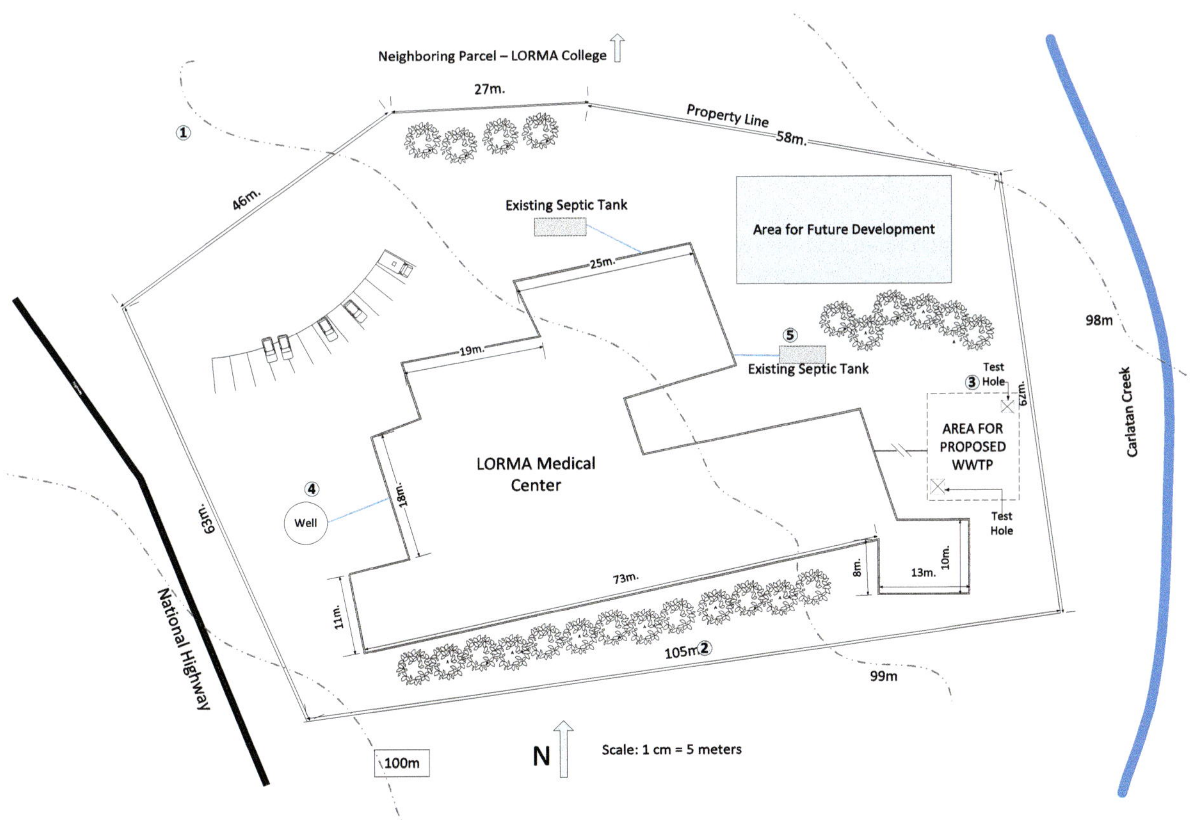

Figure 3.2 Example site plan sketch for LORMA Medical Center (Durand, 2013).

3.2.1 Step-by-step Procedure

The step-by-step process for creating a site plan sketch (which is listed in the checklist Table 3.3 below) is as follows:

(1) Choose an appropriate scale. A scale unit must be chosen that allows the plan to be best represented on paper. Often, a 1-cm increment is used to represent 1, 5 or 10 m. Proper scaling allows a reviewer to check the dimensions of the plan and the service provider to easily calculate distances and areas, which are critical in designing and sizing sewers and treatment components.

(2) Identify north on the site plan sketch and represent it by an arrow. This common engineering standard lets the reviewer visualize the site.

(3) Draw in the property lines and dimension them. Local government officials are often sources of information related to property line location. If the exact location of a properly line is unknown, a formal survey may be required. Mapping the fully dimensioned property lines helps to ensure that the DWMS is put on the correct parcel.

(4) Locate all existing buildings, driveways, roads, streams, floodplains, cut banks, and any other features of the land that might impact upon the DWMS. This is done by measuring distances between features using a tape measure and compass. Two measurements are generally required to locate a feature on the site plan.

(5) Locate any future developments, buildings, or structures. Wastewater treatment and reuse systems will become permanent features of the site. It is important to consider how the site will develop in the future, not only to appropriately plan for additional wastewater volumes, but also to make sure that future development is not limited due to poor placement of DWMS.

(6) Draw in the rough slope and topographic features. Sufficient topographic information is needed to establish the hydraulic profile which, if highly sloping, can have significant impacts on the development. Providing accurate topography will also help the service provider to:
   - select the best routes for the sewer line;
   - determine if gravity alone can make the wastewater flow as desired; and
   - determine if special features, such as sewer drops with proper drop boxes and cleanouts that are used to avoid clogging, can be installed or if pumping stations will be required.

(7) Locate all test holes, soils borings, or other soil-related testing locations. Number these so that they can be cross-referenced with soil-boring logs.

(8) Locate water lines, wells, drinking water intakes, and other water-related features. These critical features require stringent adherence to setbacks from wastewater management components to protect public health. Check with the local regulatory official or use the setbacks provided in Table 3.2 (not intended as endorsement) in conformance with local best practice.

Table 3.2 Recommended setbacks from wastewater system components (adapted from County of San Diego Department of Environmental Health, 2010).

Feature	Treatment unit (m)	Leach lines (m)	Seepage pit (m)
Structure	1.5	2.5	3
Property line	1.5	1.5	3
Property line, if wells are present	15	15	15
Public water line	3	8	8
Well	30	30	50
Cutback or drainage ditch	5	15	15
Stream or pond	10	30	30
Drinking water reservoir	20	70	70
Road easement	1	2	3

Table 3.2 provides some of the common setbacks for on-site and decentralized wastewater treatment systems. Those recommended setback distances are from the wastewater system to sensitive features of the site, and shall be used unless country-specific setbacks are provided. Setbacks from other features, such as cut banks, slopes, swimming pools, or other features should be based on site-specific considerations, such as soils and other conditions that could allow wastewater to come to the surface or otherwise impact the site feature.

(9) Locate other utilities, including electric, natural gas, phone, or cable TV lines, and their connection points. If underground utility locations are unknown, hiring a utility locating service, or carefully digging shallow holes (potholing) to identify underground utility lines is necessary, in order to avoid hazards and liability while excavating.

(10) Verify that all features are properly dimensioned. Check the distance and angles for property lines and the linear distance from the property lines to the key features of the location of the DWMS.

(11) Write in the owner name and address, parcel number, or other identifying designations. Identify the ownership of all parcels immediately surrounding the parcel in question.

(12) Provide a map or directions to the site, which will be useful for regulators, contractors, and delivery workers (Table 3.3).

Table 3.3 Site plan checklist.

Step	Y	N
Site plan sketch is drawn using a recognized engineering scale.		
North arrow is shown.		
All property lines are shown and dimensioned, and setbacks are maintained.		
Roads, driveways, water bodies, buildings, and other potentially important site features are shown.		
Future developments of the parcel are identified.		
Slopes and topography have been diagrammed.		
Locations of test holes, soil borings and soils tests are shown.		
Water lines, wells for potable water or agriculture, and intakes to public water systems are shown.		
Locations of overhead and buried utilities, and their connection points, are shown.		
Site features, property lines, and other aspects have been checked again for dimensions and angles.		
Owner name and address, and lot designation, are given.		
Information about surrounding parcels and a map or directions to the site are provided.		

3.2.2 Estimating the Amount of Land Available

Analyzing the properly-dimensioned site drawing can serve as the best method of identifying open spaces on the parcel where DWMS components may be located. The available area of those spaces can be estimated simply by using a ruler and calculator. Available land should be relatively level or as level as possible, meet all setback requirements, and be located in an area that will not inhibit future planned development of the parcel. However, where land constraints do exist, projects may be designed to use multiple areas on the parcel for DWMS. Often, treatment components such as septic tanks are located near buildings, and reuse or dispersal fields located in separate locations, often connected with long gravity (or where needed) pressure sewer systems. For this reason, identification of all potentially suitable locations is required.

Estimating land area available for the potential DWMS site(s) on a parcel

The site plan sketch is especially useful for this exercise. Once all of the features and setbacks are recorded on the sketch, the usable land area may become apparent. In some cases, the available land area will be in a square or rectangular shape. In this case, simply multiply the length by the width to calculate the available area. For more complex shapes, divide the area into different geometric shapes, such as squares, rectangles, and triangles. If your site is properly scaled, it should be simple to calculate the areas of each shape and add them together to determine total land availability.

3.2.3 Identifying site Features

Once the site plan has been drawn and checked, it is time to investigate the physical and human-related aspects of the parcel, which are grouped below.

Physical factors include the following:

- soil horizon properties, such as depth, structure, texture, and consistence;
- groundwater and surface water constraints; and
- slope and landscape position.

Human-related factors include the following:

- land use and zoning of the parcel and surrounding area;
- utility access and conflicts and any DWMS components that might already be at the site;
- site accessibility concerns;
- regulatory requirements for discharge or other regulatory issues that may influence the design or installation; and
- desired goals for the end use of the treated effluent and residuals.

The final point, understanding the different options for reusing and recovering the residuals of the treatment process, not only informs the technology decision making process, but may help reduce long term costs. Value from wastewater residuals is often not apparent, and may take a collective approach to produce enough residuals to be commercially viable.

Equipment for processing wastewater residuals (Figure 3.3) such as fecal sludge can be expensive. In agricultural areas where using treated biosolids is desirable as a soils amendment, collectives of DWMS owners may band together to purchase and operate the equipment, with revenues used to finance and operate the equipment as a business venture. Successful applications of wastewater reuse and recycling programs that are fully funded by the revenue they generate are few in number. Performing a market study prior to the purchase of equipment can help project developers better understand costs and opportunities.

Figure 3.3 Sludge drying and pelletizing machine, eThekwini, South Africa (Authors).

3.3 SOILS EVALUATION

Understanding a site's soil properties is critical in planning, designing, and installing DWMS. This is true for all DWMS, not just systems that will rely upon the soils for effluent dispersal. Knowing if the soils are shallow or deep, saturated or seasonally wet, or disturbed from previous excavations or otherwise unstable has a major impact upon technology selection and installation. This section describes the steps needed to determine if the site is suitable for installing DWMS. Obtaining this information before septic tanks and other system components are installed avoids the need for costly redesigns or use of extreme measures to install systems in less than ideal locations.

Recording the collected data on a soils discovery hole log sheet (See Appendix) provides a structured method of accessing and interpreting the data, which will help the investigator see limitations that might affect DWMS development. The following sections introduce test hole excavation and soils and slope/landscape evaluations . Additional information and case studies are provided in the Appendix.

3.3.1 Tasks Involved in Soils Evaluation

Following are the steps in the soils evaluation process. Each is introduced below and described in sufficient detail to accomplish the tasks, while more information can be found in the Appendix and references. The soils evaluation steps include:

- Excavating test holes or soils borings;
- Identifying the soils horizons;
- Evaluating the depth and thickness of each horizon;
- Identifying the characteristics of the soils that make up each horizon, by looking at each horizon's clods (example shown in Figure 3.4) for the following properties:
  - o soil texture;
  - o soil structure;
  - o soil consistence; and
  - o soil color.
- Identifying depth to limiting condition (bedrock or zones of saturation);
- Considering the site's slope and landscape position;
- Estimating the soil's long-term acceptance rate (LTAR), or the quantity of effluent that may be applied to a treatment area.

Figure 3.4 The soil texture, structure, consistence and color provide clues as to the long term acceptance rate for soils-based dispersal systems.

3.3.2 Excavating Test Holes and Soil Borings

The first step in the soils evaluation procedure is excavating soil discovery holes (test holes) or performing soil borings. This activity exposes the soil horizons and allows for testing of the different layers (strata) and observation of the depth of limiting factors such as bedrock and zones of saturation. This process is necessary for *any* DWMS that may include the installation of tanks, ponds, basins, or where soils-based dispersal fields are proposed.

Test hole excavation – process and safety considerations

- Use caution when digging test holes to prevent them from collapsing, especially if they are sited in sandy soils or in areas where groundwater is expected. Workers should wear safety apparel, such as a hard hat and proper shoes, and follow equipment safety instructions.
- Test holes should be dug to a minimum depth of 1.2 m for areas where leach trenches are proposed or 0.3 m below the minimum allowable depth of any tank hole, pond, or basin. If an impenetrable soil layer (refusal) or other limiting conditions (such as soil wetness) are met before those depths are reached, carefully recording these conditions on the test hole log form is required before one can move on to the next hole (DWMS) location. If those depths are reached without meeting a limiting condition, record the depth findings on the form and proceed to soil horizons analysis.
- For small systems, dig one test hole in the area of a proposed tank and two test holes in the area of proposed leach trenches or small ponds or basins. Excavate more test holes for larger systems or in areas where the subsurface soils conditions are believed to be variable. Try to excavate test holes to allow sunlight to reach the major sidewall during the observation period, so that subtle differences in soil color will be easily visible.
- If test holes are to remain open overnight, protect them with fencing, or at the very least mark them with caution tape or other barriers to keep out people and animals.
- Ramp one end of each test hole to allow for entry for close evaluation of the soils strata. Workers should never enter a test hole to a depth greater than 1.25 m below the ground surface unless the test hole is protected by shoring (stabilizing the sidewall of the hole by installing wooden or steel supports) or excavated with sloping sidewalls. Work around the hole rim should be done so as to prevent cave-in, especially if soils are unstable, or if soils contain loose sand or gravel. Heavy objects should not rest on or near the test hole opening.
- Fill in test holes after the last inspection so that they will not pose a hazard.

3.3.3 Identifying and Evaluating Soils Horizons

Soils contain particles of different sizes that combine together into soils layers (horizons) with distinct structures and textures (Figure 3.5). These qualities influence the ability to transmit water, quantified as the long-term acceptance rate (LTAR). The horizons are developed over time through the action of wetting and drying, plant root penetration, freezing and thawing (in certain climates), and mineral weathering.

Figure 3.5 (a & b) Soil horizons are often distinctive in terms of color and texture and are easily identified (Lindbo, 2013).

Soils evaluation requires identifying the separate horizons that are present within the test hole. This allows for the classification of the horizons' properties that influences LTAR and other concerns for DWMS application. The classification describes each horizon's depth and thickness, texture, structure, consistence, and color, each of which is described below.

3.3.3.1 Soil Depth and Thickness

Determining the depth and thickness of a soils horizon is a simple visual process. Soils horizons are nearly parallel to the surface, with obvious boundaries and differences in color and texture seen by scanning the test hole(s). Scraping away loose particles with a screwdriver, knife, or trowel makes these differences even clearer. Once the horizons are identified, their depth and thickness are then measured against a fixed point using a standard tape measure. Those values are recorded and the horizons are then used to guide the following analyses, which are done for each horizon and listed and filled in separately in a log form (in the Appendix) used for formally describing the site and estimating the appropriate LTAR.

3.3.3.2 Soil Texture

Soil texture refers to the sandy, silty or clayey nature of the soil, which influences how fast water can pass through the soil (its permeability). Therefore, texture is based on the composition of sand, silt, and clay particles in a given sample. Making a correct assessment of soil texture is critical as this feature determines the baseline LTAR that will be adjusted by the other features described below.

Texture is most accurately found by passing soils through sieves with different mesh sizes that separate particles by size into three categories: 0.05 to 2.0 mm for sand, 0.002 to 0.05 mm for silt, and <0.002 mm for clay. Testing many soil samples in the horizon helps users determine what percent of the horizon is sand, silt or clay. Those values are then used to trace the gridlines in the Soil Texture Triangle (Figure 3.6) to find the overall horizon's texture.

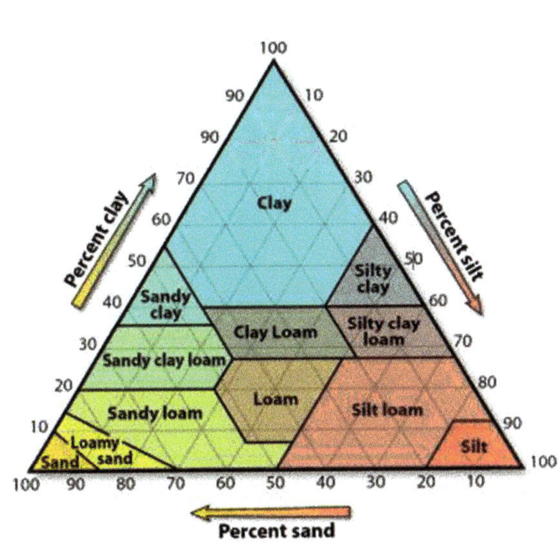

Figure 3.6 The Soil Texture Triangle (Lindbo, 2009).

An alternative method, which is easier because it does not require using sieves, evaluates texture by feeling a wetted sample of soil and judging its texture based on how:

- sand feels gritty between the fingers;
- silt feels like talc or flour; and
- moist clay can be forced into long ribbons between the fingers.

With practice, this simple technique can yield fairly good results. A more advanced sensorial method (Figure 3.7) that can improve accuracy simply involves wetting the soil, forcing it into a ribbon, and evaluating the length and smoothness of the ribbon to determine the texture of that soil sample. With a little practice, one can become skilled at quickly determining soil texture in the field. The sensorial soils evaluation processes along with other guidance suggestions are fully described in the resources listed within the references at the end of this chapter and in the Appendix.

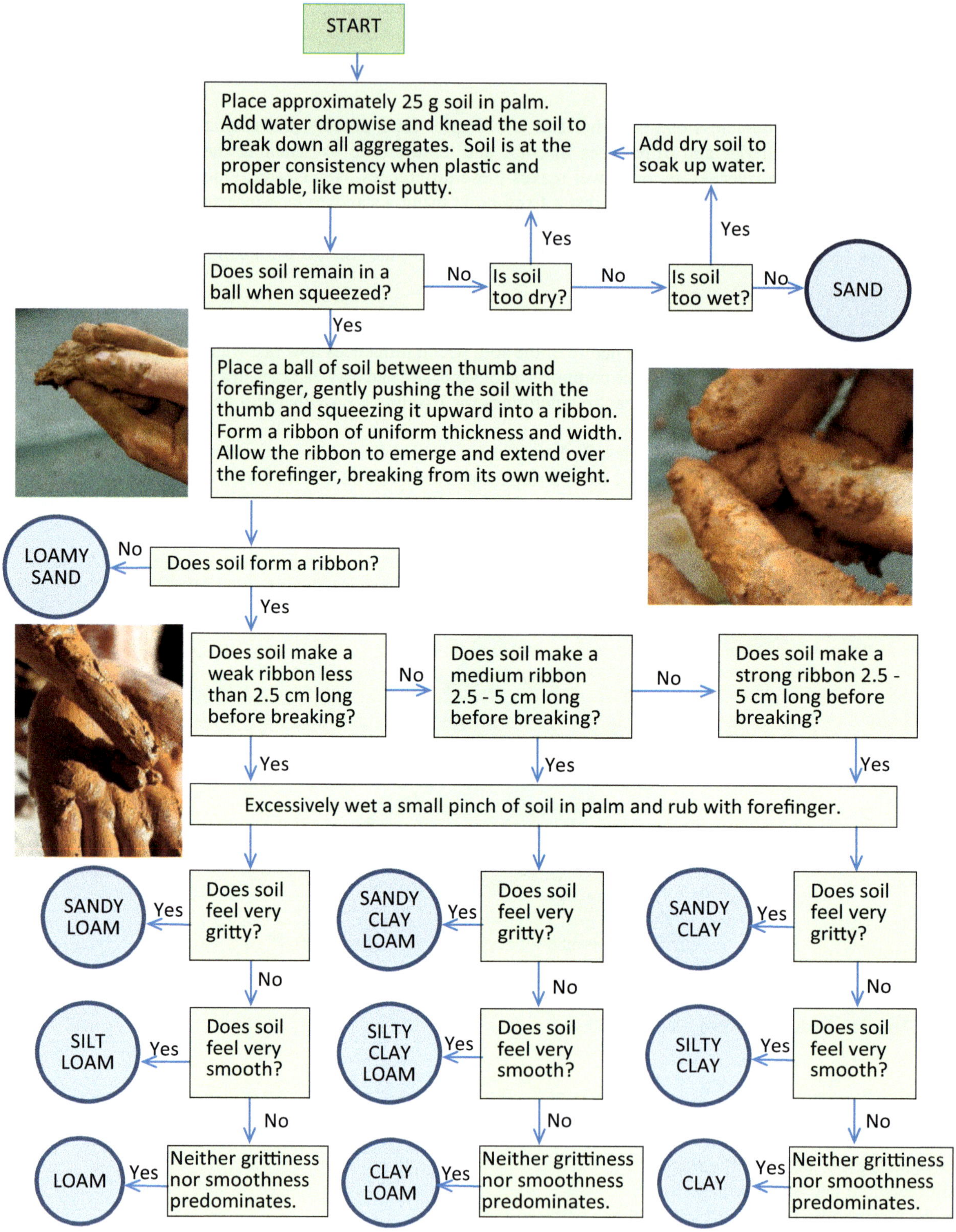

Figure 3.7 Advanced sensorial soils texture evaluation (adapted from Lindbo *et al.*, 2005).

The relationship between soils horizons, the presence or absence of abrupt texture changes, and the amount of compaction within the soil horizon may have greater influence on the ultimate LTAR of the soil sample than does the texture itself. Simply put, knowing texture is not enough to adequately design a system or predict how that system will function over time. It simply establishes a possible range of LTARs to use for design purposes. The other aspects of the soils evaluation described below will help to better make this determination.

3.3.3.3 Evaluating Soil Structure

While soil texture describes the percentages of sand, silt, and clay, soil structure describes how these particles are arranged in larger soil formations known as peds. The grade, size, and type of soil peds, described below, affect how water will move through soils, meaning that like texture, soil structure relates directly to how well soils will disperse and treat wastewater over the long term (and therefore factors into the LTAR calculation). The three components of soil structure are:

(1) grade, or how strongly the soil structure is expressed (degree of random arrangement);
(2) ped size, ranging from very fine to very coarse; and
(3) ped type, or shape (shown in Figure 3.8).

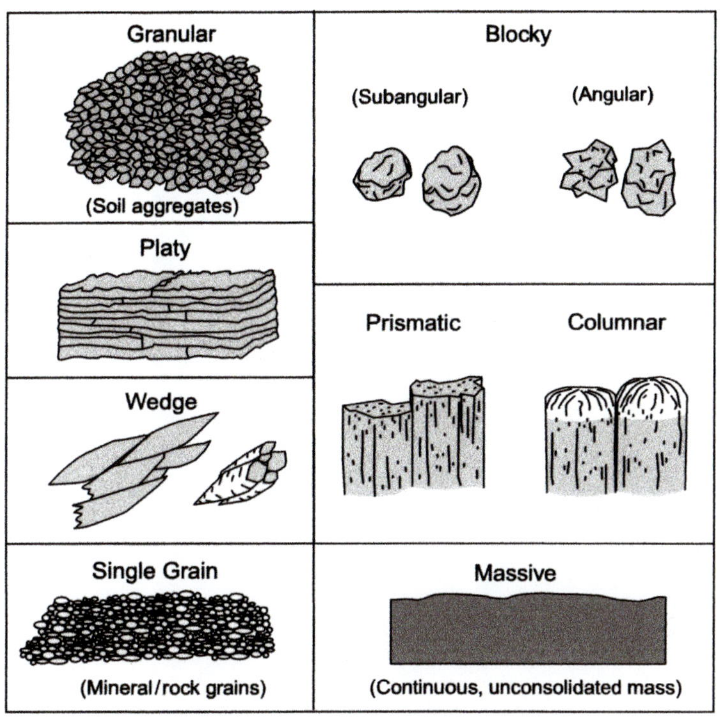

Figure 3.8 Common soil structure ped types (Lindbo, 2013).

Soils structures take years to form and can be easily disturbed or destroyed. This can have a significant impact upon how well the soils will function in a DWMS over time. For this reason, soils-based dispersal systems installed in sites that have been previously disturbed must be properly designed with this in mind. Poor construction practices during the installation process that might destroy the soil structure, such as excavating during excessively wet seasons, should be avoided. Soil structure and the other parameters that are studied within the full methodology for classifying soils are all described in the Appendix.

3.3.3.4 Soil Consistence

Soil consistence describes how easy it is to deform or break up the individual soil peds. It measures the soils' presence of clay particles that may shrink or swell during wetting, and how likely the soil may be to smear during installation. Both conditions can negatively affect the long-term performance of the system. Like with the other soil properties, more information on soil consistence and how it factors into the LTAR calculation can be found in the Appendix.

3.3.4 Identifying the Water Table and Other Limiting Conditions

To protect public health and ensure the long-term operation of soil dispersal systems, discharging wastewater effluent into saturated soils is prohibited. If the water table meets the bottom of a leach trench, a clogging zone may form more rapidly, leading to early system failure and a direct connection between soils-based dispersal systems and drinking water aquifers.

Therefore, it is necessary to understand the three states of soil saturation, which are as follows:

- *unsaturated* – soils with pore spaces between soil particles that are filled with air or water that is transient;
- *saturated* – soils with pore spaces between the soil particles that are always full of water. The elevation of the top of the saturated zone is known as the water table;
- *seasonally saturated* – soils that are flooded seasonally, usually during rainy seasons when the elevation of the water table rises or for any period of time longer than 7 days.

Often, the elevation of the uppermost level of saturated soil changes during the year. From a site evaluation perspective, the goal is to identify the depth to the seasonal high water table (SHWT). There are two good ways to identify the SHWT. The most common method is to observe mottles, which are areas of red and grey discoloration of the soil due to the reduction, translocation, and oxidation of iron oxides in saturated and anaerobic soil conditions (Figure 3.9a). In general, finding the SHWT can be done at a basic level by assessing the horizon depths for mottling, or rigorously through using the Munsell Chart color analysis procedure as described below.

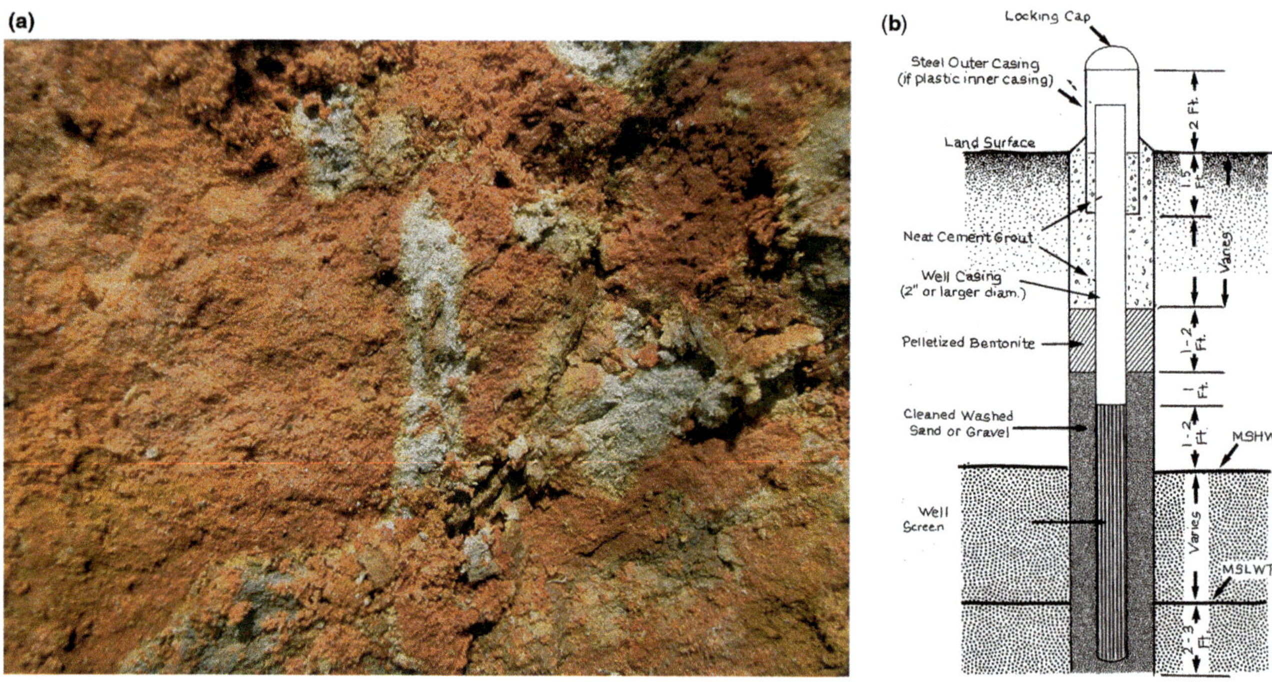

Figure 3.9 Soil mottling as an indication of soil saturation (Lindbo, 2013); shallow groundwater monitoring well (Authors).

The second method is by monitoring water levels within wells in the area. However, with cased wells, shallow water is often sealed to encourage withdrawal of deep water. In these situations, consider installing shallow groundwater monitoring wells (Figure 3.9b). Water depth then can be obtained using a probe or with a tape measure and flashlight. Note that monitoring for water table elevation must occur during the times of the year when the highest groundwater levels are expected.

The Munsell soil color chart helps with detailed recognition of soil mottling and is advisable for larger soils-based dispersal system projects as an accurate method of determining depth to seasonal high water. For more information, refer to the U.S. Illinois Soil Classifier Association (2010) soil investigation resource: http://www.illinoissoils.org/Links_Files/Understanding%20Soils_Final2.pdf.

3.3.4.1 Depth to Limiting Conditions

Limiting conditions are soil features that make a site unsuitable for dispersal either based on texture, structure, bedrock (totally impermeable) status, or saturation. Often, limiting conditions are identified by simply observing standing groundwater, significant mottling as described above, impermeable layers that may be associated with dense clay, or the presence of bedrock. If no such features are obvious, then the limiting condition can be chosen as the horizon with the highest percentage of clay. In any case, knowing the depth to the shallowest limiting condition, which should be separated from the bottom of the leaching system by at least 0.3 m of usable soil to ensure proper treatment, will help determine the maximum potential depth of soils-based dispersal systems. A complete description of the how that depth can be estimated and how it fits into the soil dispersal system sizing procedure is provided in the Appendix.

3.3.5 Evaluating Slope and Topography

Moving up to the soil surface, the focus now turns to slope and topography. Slope defines the incline between two points. Topography reflects the surface shape and features of an area, which form the basis of the slope calculation as well as overall landscape position. Both are important for evaluating sites intended for DWMS.

3.3.5.1 Assessing Slope and its Impact on DWMS

Slope can be expressed in two different ways – as a percent incline or as an angle of the slope. The initial steps to calculating slope either way are the same. The basic calculation of slope is shown in Figure 3.10. Often called rise over run, this refers to vertical distance divided by horizontal distance, which is multiplied by 100 to give a percentage.

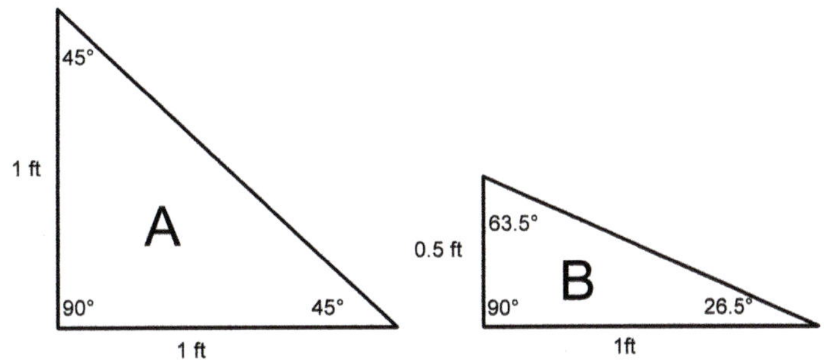

% slope = (rise ÷ run) x 100
Triangle A slope = (1ft/1ft) x 100 = 100%
Triangle B slope = (0.5ft/1ft) x 100 = 50%
45° slope = 100% slope
22.5° slope = 50% slope

Figure 3.10 Examples of slope calculation (Lindbo, 2009).

The slope of the land is strongly linked to the technology and cost of DWMS installation. DWMS may be successfully installed on slopes as high as 60 percent, although the amount of fill needed and excavation issues may make finding less-sloped sites (less than 30% slope) more attractive. Slope also has an impact upon the design of soils-based dispersal systems, as shown in later calculations.

3.3.5.2 Topography's Relationship with Slope

Slope also can be calculated easily by using a detailed topographic map (Figure 3.11) if one is available. This common task is often performed when considering routes for sewer lines. Once the sewer line location is plotted on the map, the length of the line (distance) must then be measured, and using the scale of the map, converted from feet to meters if necessary. The total elevation change along the line can then be determined by subtracting the elevation of the lowest contour used from the elevation from that of the highest contour used. No conversions for this measurement are necessary, as it is a real-world elevation change.

To calculate a percent slope, simply divide the elevation change in meters by the distance of the line drawn and multiply by 100 as instructed above. For example, if the value calculated is 20, then for every 100 meters covered in a horizontal direction there will be a gain (or loss) of 20 m in elevation.

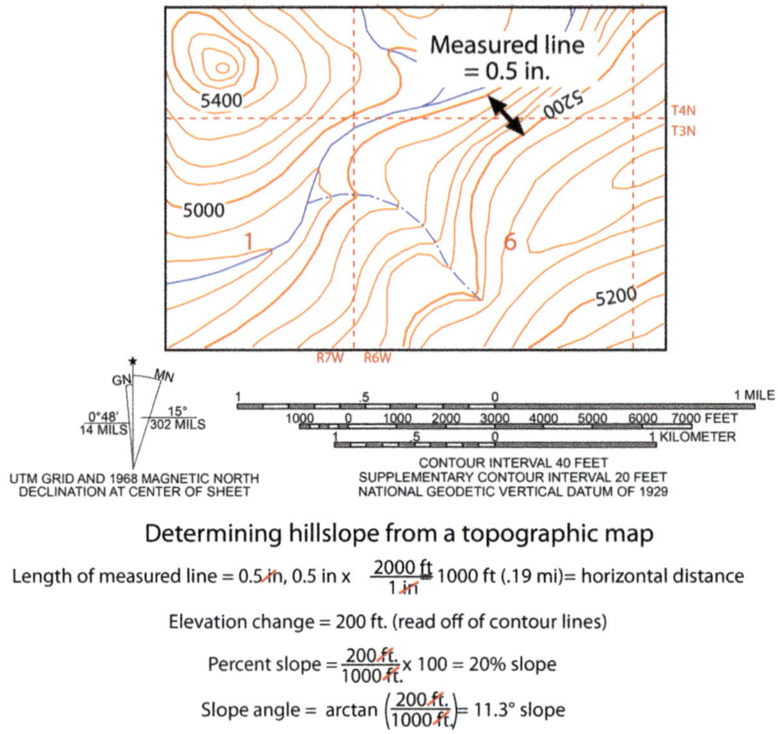

Figure 3.11 Using topographic maps to determine slope (Geospatial Training and Analysis Cooperative at Iowa State University [GSTAC], 2008).

To calculate the angle of the slope, divide the elevation change in meters by the distance of the drawn line (after converting it to meters). This is the tangent value for the angle of the slope. Apply an arctangent function to this value to obtain the angle of the slope, by hitting the 'inv' button and then the 'tan' button on most scientific calculators to get the slope angle. This will provide the slope angle calculated between a horizontal plane and the surface of the hill. Even if you do use a contour map, be sure to verify all elevations in the field.

Determining topography in the field

To verify elevations in the field, a simple builder's level, or even a water-filled tube level, can be used to make very accurate slope and basic topographic assessments. One strategy is to lay out a grid by placing stakes at evenly spaced intervals (points) on the site. Using more stake-points increases the accuracy of the topographic determination. Once the stakes are placed, measure the elevation change using a builder's level from a fixed elevation point. Plot these elevations on the site plan and connect points with common elevations to indicate the contours.

For more complex projects, hire a land surveyor who will use more accurate methods to develop topographic maps and identify property lines. This is useful because often there is no topographic information (much less a map) for a site, and it will be up to the service provider to produce one.

3.3.5.3 Landscape Position and Slope Description

Beyond the scale of point-to-point slope, topography also shapes the landscape position of a site, which also plays a large part in determining a site's suitability for soils dispersal-based DWMS. The most notable concept within landscape position is that water will tend to pool, increasing soil wetness, within areas with concave (bowl-shaped) or flat features. Wetter soil conditions imply increased ease of potential pathogen passage through the soil to the groundwater, as well as poor drainage that can lead to storm water or wastewater pooling up at the DWMS site and harming the system. Therefore, toe slope and near-stream zones (often classified as floodplains), as the most concave zones, are generally considered unsuitable for

DWMS. However, soils are often very shallow at the ridge top (head), which can sometimes make the ridge top areas also unsuitable for DWMS.

The interrelated concept of slope description, which defines a site by its concave, convex (opposite to concave) and linear nature both in the downhill and horizontal directions, also influences the calculation of the site's LTAR. The methodology for classifying slope description as well as landscape position is better described in the Appendix, which also shows how the two parameters can be used within LTAR calculations.

3.3.6 Determining Long Term Acceptance Rate and Identifying Other Soil-based Concerns

As mentioned previously, calculating the LTAR is a function of a site's soil texture, structure, and consistence; slope and landscape position; and vegetation (which can increase permeability) and other factors. Refer to the Appendix for a full description of this topic and an in-depth review of the soils evaluation procedures. Loading rate calculation forms and examples are also provided.

As for other soil-related issues, understanding the buoyancy forces at work if soil saturation is present at a given site is critical. Buoyancy, as it relates to DWMS, is the upward pressure exerted by groundwater that opposes the weight of a tank or other structures. Buoyancy forces are powerful and can lift concrete septic tanks out of the ground if they are not properly installed (Figure 3.12). In practical terms, the buoyancy force equals the weight of the displaced water by the tank or structure. Designers should ensure that there is enough weight (the weight of the tank and soil cover on top of the tank, or concrete substitution for soil cover) to counterbalance the buoyancy forces of the tank when empty.

Figure 3.12 (a) Septic tank disaster due to buoyancy forces (Lindbo, 2013); (b) Using tie-down to secure septic tank to concrete slab in high groundwater conditions (Authors).

> **Septic tank anti-buoyancy force design considerations**
>
> Septic tanks are almost always full of water, so buoyant forces are mostly canceled out. However, when tanks are desludged, they may float if groundwater is present. For this reason, service providers must consider buoyancy forces when designing DWMS in saturated soils. Each liter of air in the tank will try to lift the tank with a force equivalent to the downward force of a stationary 1 kg object. So, a 5,000 L empty tank in a water table zone has to weigh at least 5,000 kg or be tied down securely; otherwise, it may float.
>
> Attaching the septic tank to a concrete foundation is one solution. The weight (mass) of the soil on top of the tank and the weight of the tank itself should also be considered when counteracting buoyant forces.
>
> For more information, view this sample calculation sheet (in non-metric units) from Roth USA (2007), an environmental services provider: http://www.roth-usa.com/PDF_Download_Files/RMT_buoyancy_calc.pdf.

3.4 DETERMINING IF SURFACE WATERS ARE LIMITING CONDITIONS

In developing countries such as Madagascar, Bangladesh and Indonesia, large populations of low-income or displaced people are often forced to live in flood-prone areas where space is most readily available. This not only places people at risk from the hazards brought by floods, but pollutes surface waters due to flooded sanitation systems' release of un- or under-treated human wastewater. Therefore, analyzing surface waters and how they may impact a parcel of land chosen for DWMS is important for site evaluation and for selecting appropriate technologies. Below are definitions helpful to this discussion:

- *Floodplain* – the flat land bordering a river, which stretches from the banks of the channel of the river to the base of the valley walls, that experiences flooding during major storm events.
- *Floodway* – the channel of the river or stream that carries flood flows.
- *Flood fringe* – the land that may be covered by flood waters during storm events or periods of high discharge, but that does not experience strong currents.
- *Flood elevation* – the vertical elevation that water levels reach during floods. Often referred to as 10-year, 50-year, or 100-year flood elevations, the period designates the likely highest elevation expected to be equaled or exceeded during the period.

DWMS components should never be placed in a floodway and only be placed in a floodplain if full controls are implemented, such as:

- properly designed riprap (bank stabilization) systems that shield equipment from flood flows;
- full protection from inundation with watertight covers and access ports; and
- components elevated above the 100-year flood elevation and installed in waterproof electrical boxes.

DWMS components may be installed in a flood fringe area as long as partial controls are installed. Typically these are watertight tank access ports, lids, and electronic equipment, which must be installed above the 100-year flood elevation.

Information about flooding can be obtained using floodplain maps, published information, or anecdotal information. Unfortunately, accurate floodplain maps are limited or nonexistent in many developing countries. Due to this lack of published data, it is often necessary to consult neighbors or historians who have firsthand knowledge of flood elevations based on past significant flood events. In areas of the world that are forecasted to experience wetter conditions or more severe storms due to climate change, designing DWMS to withstand estimated future high water levels adds resiliency.

3.5 DETERMINING HOW LAND USE OF SURROUNDING PARCELS AFFECTS DWMS SELECTION

The land use activities on neighboring parcels, which are often shown on planning maps (Figure 3.13), may affect the wastewater treatment technologies that are appropriate for a project. In the case of the city in the figure, dark green areas (which are farmland) would permit neighboring DWMS such as sewage lagoons that would likely be different than those DWMS permitted on nearby red areas (new development). Overall, it is important to know about the land use patterns neighboring the system for the following reasons:

- neighboring developments may be able to connect to the system in the future, which could provide additional revenue from fees. Knowing this in advance can help planning for future expansion;
- neighboring land use and water-related features may impact a site's technology decisions in more basic ways. For example, wells on neighbors' properties need sufficient setbacks from DWMS just like the ones on the study site do. Also, if there is a population center next to the DWMS site, technologies that generate odors and/or noise, such as sewage lagoons, should be avoided or managed with nuisance control measures. Use Table 3.4 as guidance.

The main issues as seen in Table 3.4 are odor, insects, and noise. These are more of a problem for larger DWMS systems, such as those serving subdivisions or large commercial facilities, rather than small systems or those serving single-family residences.

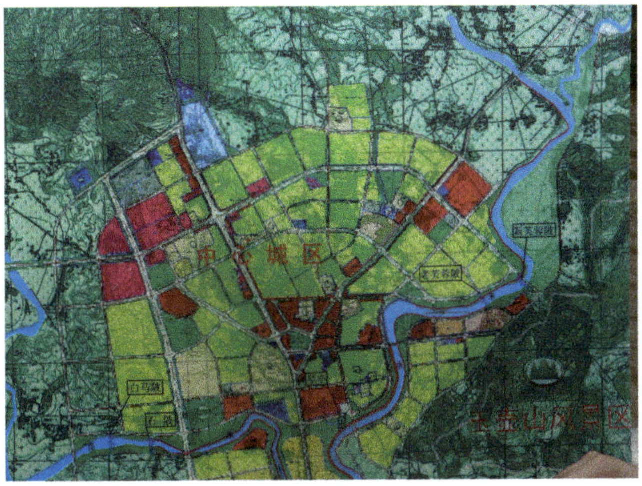

Figure 3.13 Land use planning map for Pingxiang Municipality, China (Authors).

Table 3.4 Odor and noise control recommended measures for larger DWMS systems and examples based on distance from property lines.

Distance from wastewater system to property line	Nuisance control measures	Reason(s)	Examples of control mechanisms
50 m or less	Full odor, noise, and insect controls.	At this distance, odors, and noise from wastewater systems can cause complaints.	Gas-tight covers on ponds; gas-tight seals on all manholes and riser lids; proper insulation on mechanized equipment. Full enclosure is an option.
50 m to 100 m	Partial odor control and insect control.	Odor can travel over the ground surface for long distances if winds are right.	Gas-tight covers on ponds; gas tight seals on all manholes and riser lids.
100 m and above	Insect control.	Flies and mosquitoes have large ranges. Insect control should always be practiced.	Good housekeeping practices; mosquito-control measures.

Customizing plans to avoid DWMS nuisances based on land use

The wide variety of land-use definitions and related requirements mean that site-specific analyses and solutions should take the above nuisance-control concepts and measures and apply them as needed. This process should always start by investigating nearby land uses and the governing regulations. Planning and engineering offices for local governments are good places to start, as zoning and land use maps may be available. Those offices can also give information about local projects and experiences in DWMS nuisance control, which will help with understanding local conditions and regulations. This is important, because odor and noise controls may be expensive, and using inappropriate controls for the context and extent of the nuisance may be a waste of resources.

3.6 IDENTIFYING UTILITIES, PRE-EXISTING DWMS, AND RELATED CONFLICTS

Determining which utilities pass through a parcel is a required step in site evaluation. It helps avoid accidentally breaking buried utility lines and can serve as a reminder about the hazards associated with overhead wires and heavy equipment such as backhoes. Also, this process gives useful information in the case that systems are mechanized and will need electrical power and/or water supply (which is often needed for installation or O&M). Finally, other concerns include understanding if or how pre-existing DWMS components will affect the project, and if such DWMS might be usable or need to be removed.

3.6.1 Electrical Utilities

Only licensed electricians may perform work on the electrical components of DWMS. However, the property owner also may need to get involved by working with local utilities to determine if power lines may be accessed.

> **Electricity considerations**
>
> Electrical power can be single phase or three phase with voltages ranging from 120 V up to 440 V in common applications. Motors and other electrical equipment must be matched with the proper phase and voltage, which should be done with the assistance of the local electric utility. Also, just because there is a power line crossing the property, do not assume that it may be tapped into. Avoid expensive project revisions and find out about power availability during the site evaluation stage.

3.6.2 Water Availability for System Operation and Maintenance

Unlike with electrical lines, which are usually above-ground, it is necessary to find out where underground water lines are located. If the property owner is unsure, cautiously dig shallow exploratory holes (potholes) before excavating. Water lines may be installed anywhere, and they are easily broken.

> **Water access considerations**
>
> For many DWMS configurations, water availability, such as a water tap from a piped municipal system that is conveniently located near the system, is a benefit. Water supplies are often needed in the construction or operations process, such as when leak-testing a septic tank. Water availability may be an important factor in making technology decisions. Also, water line location, as discussed above, is an important consideration to insure proper setbacks from DWMS components.

3.6.3 Existing DWMS Components

Existing or abandoned DWMS equipment may be used in new or expanded construction. Existing septic tanks may have several uses in system configurations, and incorporating them may save money. Before considering if this is advisable, the tanks first need to be found. Then, the service provider can perform a simple leak test. Leak tests must be performed for new construction as well, as ensuring that and outlet ports are sealed and leak tested prior to backfilling makes leak identification and repair easier.

Determine the usability of the tank by evaluating it for proper sizing and soundness of construction. To verify the size, uncover the entire tank surface using probes and estimate the volume (length times width times depth). The depth is measured from the bottom of the flow line of the outlet pipe. As for soundness, the proper access ports, risers, and baffles need to be in place (and for the first two, sealed) for the assessment, while inlet and outlet piping must be verified to be suitable for continued use.

> **Dealing with leaching septic tanks**
>
> In developing country settings, septic tanks often are designed with leaching compartments that discharge the settled effluent through a leaching chamber, which is an attached open-bottomed compartment of the tank. This is an improper design that can result in failure or groundwater contamination. The discussion above on soils suitability and permeability testing should indicate that for proper DWMS, leaching septic tanks should not be permitted.

To assess if an existing septic tank has an illegal leaching chamber, first pump the tank and remove any remaining sludge that has accumulated at the bottom. Visually inspect for gaps, openings, or voids that may indicate non-watertight conditions. If a leaching chamber is discovered in an existing tank, it may still be used for new DWMS as long as the leaching compartment is bypassed by piping directly from the primary settling chamber of the tank. Also, existing tanks with leaching compartments can be sealed with plastic liner. Both procedures are difficult tasks that may not be worth the effort or save money in the long run.

3.6.3.1 Accessing Existing Septic Tanks

To determine the accessibility and usability of existing septic tanks, it is necessary to first find them. Often, the precise location of the septic tank and/or its condition is unknown. If this is the case, follow these procedures:

(1) Locate the septic tank. The tank should be approximately 3 m from the foundation of the building, but often, in space-constrained urban settings, the tank may be installed directly under the kitchen. The drain pipes leading out from the home or under the home should give you a good indication of the tank's general location. Depressions or raised areas around the home may also hint at the locations of tanks.
(2) Dig down to the tank. In most cases the tank is no more than 0.5 m below the surface. Once the tank is reached, locate the access port on the top of the tank.
(3) Clear away enough dirt to allow for liquid removal through the access port. The lid may be quite heavy, so the owner may need help when removing it. Once the lid is removed, insert the pumping hose into the tank to remove the sludge.
(4) Breaking a hole in the top of the tank may be the only option available to access the tank for desludging if there is no access port. This can often be accomplished with metal bars and hand tools. Upon completion, installation of a gas-tight riser and locking lid will be required and will provide easy access in the future.
(5) Once this process is completed, fill the hole and clean up.

Other considerations for locating septic tanks

In many instances, septic tanks are installed under the home, either beneath the kitchen or the toilet area. To locate them, use a metal bar and gently tap the floor (the tic-tic-tic method). Septic tanks will make a hollow sound when tapped upon in this fashion. While it is less desirable to excavate a hole in someone's kitchen floor, if no other access port is found, this will be required. After desludging, the tank should be fitted with a gas-tight lid that will permit easy access during the next desludging event.

3.6.3.2 Leak Tests – a Step-by-step Procedure

Even if the tank passes a visual inspection of size and internal soundness, checking the watertightness of the tank by using a leak test is still required (Figure 3.14). While all new septic tanks, sewer lines, and basins should be leak-tested after they are installed, this step must be done as part of the site evaluation if existing DWMS infrastructure is being considered for usage. To do so, follow the simple steps below:

(1) seal inlet and outlet pipes as well as risers using silicone caulk or other suitable material that will not shrink;
(2) fill the tank with water to an elevation 10 cm above the flow line of the tank or basin or any connecting pipe or structure;
(3) allow it to sit overnight (12 hours minimum);
(4) refill the tank or basin as needed to bring the water level back up to 10 cm above the flow line or level of any protruding pipe or structure;
(5) after refilling, measure the water level exactly from a fixed point (the top of the tank or a mark placed on the tank wall);
(6) measure the water level drop after 1 hour. For septic tanks, if there is more than a 0.5 cm drop in water level in 1 hour, drain the tank and repair. For small seeps, add 1 kg of Portland cement for every 1,000 L of water to the tank water and keep it in suspension by stirring with a paddle for 4 hours. As the water seeps out, the Portland cement particles will fill the hole; and
(7) after sealing, repeat the leak test to verify. For persistent leaks, drain the tank and allow it to dry. Then seal it with waterproof cement or approved waterproofing compound. As a last resort, remove the contents of the tank and install a PVC membrane in the tank. PVC membrane is available at plumbing supply outlets.

Figure 3.14 ABR preparation for leak test (Authors). Workers have sealed the outlet hole and will conduct the leak test prior to installing the tank lid, which will make any needed repairs easier.

Leak test considerations

Prior to leak testing the septic tank, ensure that all risers and cleanouts are installed to the proper grade and sealed with grout, or silicone caulk that will not shrink (non-shrink caulk). The water level for the leak test should be at 10 cm above the highest level where a pipe or riser enters or attaches to the tank.

Leak tests should be performed on new septic tanks, other tanks, basins, sewers, and other wastewater-holding components. Facility owners that subcontract DWMS construction should require leak testing and third-party verification as a condition of final construction acceptance.

3.6.3.3 Removing or Abandoning Existing Septic Tanks

In some instances, such as when new sewer systems make existing septic tanks unnecessary or a tank has serious leakage, the tanks should be abandoned. Septic tanks may be abandoned by either removing the tank or abandoning it in place.

3.6.3.3.1 Removing Existing Septic Tanks

The contents of an existing septic tank may be hazardous even if it has been sitting unused for several years. For health and safety reasons, remove the contents of the tank and disinfect it prior to removal. To remove an existing septic tank, follow these instructions:

(1) Have a licensed service provider pump the tank to remove as much of the liquids and settled sludge as possible;
(2) Add water and use a pole or other device break up any remaining sludge and then pump again, taking care to remove as much of the liquid as possible;
(3) For plastic or fiberglass tanks, it may be possible to remove them without breaking the structure down into smaller pieces. For concrete tanks, using heavy equipment like a backhoe (to break the tank into smaller pieces), jackhammers, or even manual labor to break the tank will likely be required. Use caution and personal protective equipment to protect workers from exposure to any remaining pathogens; and
(4) Dispose of tank debris in a proper landfill.

3.6.3.3.2 Abandoning Septic Tanks in Place

To abandon a septic tank in place, follow these instructions:

(1) Pump the septic tank, taking care to remove as much of the liquid and accumulated sludge as possible;
(2) Break a hole in the bottom of each compartment of the tank using heavy equipment. Never enter the tank to accomplish this task;
(3) Fill the septic tank with sand so that all void spaces are filled; and then
(4) Cover with native soil and compact.

3.7 EVALUATING ACCESSIBILITY ISSUES FOR DWMS

DWMS should be planned so that periodic desludging, inspection, O&M, and repair or replacement can occur and so that system components are safely secured from accidental damage or vandalism. Therefore, the proposed site's geography and vehicle access to the premises, which improves the ease of system service, may affect technology selection. Evaluate system access to ensure that materials may be safely and efficiently delivered for the system construction and that access is open for all inspection and O&M personnel.

3.7.1 Assessing Site Accessibility

Sites where DWMS are installed must be accessible to workers and regulatory staff. Assess access for:

- equipment and materials delivery. Consider weight restrictions for certain DWMS technologies that require bringing heavy loads of sand, gravel, or rock to the site by dump truck;
- the desludging truck. Septic tanks or other components that need desludging require easy access for desludging equipment. Most desludging vacuum trucks have standard 25 m hoses, and are limited by vertical lift of 4 m (the distance from the bottom of the tank to the inlet elevation of the tanker truck). If these distances cannot be met and access is limited in general, identify other options for desludging (Figure 3.15). Contact the local desludging service provider to understand which options may apply to the project;
- O&M personnel. Once the system is installed, access must remain open for service providers so that they may perform routine and emergency service tasks as needed; and
- inspections. Often, regulatory inspectors will require site access for routine inspections or sampling to verify compliance.

Figure 3.15 Example of manual desludging operation (Authors). Workers have sealed the outlet hole and will conduct the leak test prior to installing the tank lid, which will make any needed repairs easier.

While system security is important for larger municipal wastewater systems, it is also a concern for DWMS. Security needs to be addressed during the full site evaluation process. Wastewater systems can attract unwanted attention, and keeping the public from accessing (or vandalizing) them is important. Signs that warn of the dangers associated with the system, security fences, and locking manhole lids protect the public from the inherent dangers associated with wastewater systems. Knowing the costs of access and security measures from the beginning ensures that they are included in the budget.

Ensuring DWMS access and protection in Georgetown, Guyana

In Georgetown, Guyana, the Partners of the Americas Farmer to Farmer program's St. Stanislaus training farm runs a model cow waste biodigester program that provides training and capacity building for proper DWMS management. A key topic at the center is site safety, accessibility, and security. These fundamental concepts in wastewater system design help ensure ease of operation and protect the system from damage. In rural settings, securing the wastewater treatment site from animals is necessary for the longevity of the system. Not only do proper access controls allow for inspections, but it also prevents fluke accidents (such as a cow trampling on and rupturing the tube) or exposure to sunlight or flooding that can degrade system components. As shown in the left image (of a staff member inspecting the gas regulation system) in Figure 3.16, this requires simple roofing and flexible entry fencing, which at the right of the manure input drum can be tied in place or lifted back for easy entry.

Figure 3.16 Tubular biodigesters showing proper (at left) and non-existent (at right) access and protection measures in Guyana (Authors).

In contrast, outside of the city, a small-holder's tubular digester system was not sufficiently protected, which contributed to temporary system failure, as seen in the right image of a temporarily deflated digester. As shown, sunlight can directly strike the plastic; rain can collect on the digester and soften the soil beneath the unit, allowing the foundation to shift and affect slope; and chickens can walk on the unit and potentially peck it, leading to deflation and the need for avoidable repairs. All of this could have been prevented by simple fencing and roofing. This goes to show that proper access and protection are crucial to determining system success. While this is especially true in the case of digesters on farms where animals can damage systems, it is also true for more reinforced DWMS technologies that can still require manual O&M on certain components and can still be damaged by sunlight or flooding innocent children or interested community members (or even vandals) who can touch and break key system parts.

3.8 BENEFITTING FROM WASTEWATER RESIDUALS

Reclaiming, recycling, and reusing wastewater treatment residuals, such as treated effluent, biosolids, and biogas from anaerobic digestion, can have broad appeal if the site or nearby parcels or markets can cost-effectively benefit from using them. The residuals may be an important incentive that can encourage DWMS investment. Residuals can be defined as follows:

- treated effluent that may be used for landscape or crop irrigation, stall flushing for livestock operations, or toilet-flushing;
- grease from grease traps that may be useful as a feedstock for biodiesel or other commercial products, or as a fuel source;
- biogas from the anaerobic digestion process that might be used for cooking or heating; and
- biosolids that might be composted and used as soils amendment for use on agricultural crops.

Sites for potential opportunities for reusing effluent or wastewater residuals for various activities may include:

- gardens or agricultural plots that may benefit from effluent irrigation;
- landscape areas where biosolids application and/or effluent irrigation might add value;
- pond sites that might be favorable for effluent-enhanced aquaculture or biomass harvesting that also serves as additional treatment;
- nearby buildings that would benefit from properly treated and plumbed effluent for toilet flushing; and
- facilities that might benefit from methane for cooking, heating, or commercial processes.

These potential reuse activities can add value and/or reduce the overall costs of DWMS. Therefore, prioritizing the end use of the residuals has a large impact upon technology selection. Use Table 3.5 for guidance.

Table 3.5 Reuse measures, concerns, and recommended technology solutions.

Reuse activity	Technology level	Major related concerns	Possible system configuration
Surface irrigation of food crops	Primary and secondary treatment, tertiary filtration and disinfection	Potential contact with or ingestion of pathogens by agricultural workers and consumers	ABR, constructed wetlands, slow sand filtration, and chlorine or UV disinfection
Subsurface irrigation of non-food crops	Primary and secondary treatment	Balancing crop needs with availability of the effluent, and nutrient content.	ABR, constructed wetlands
Composting biosolids	Systems that collect, process, and dewater sludge	Potential contact with pathogens; keeping proper carbon to nitrogen ratio, moisture content, mixing scenario and temperature	Pit toilet, septic tank or ABR located at the house or structure; lagoon or biodigester for sludge stabilization, and constructed wetlands or biofilter for effluent treatment
Biogas reuse	Anaerobic digestion, secondary treatment	Gas collection safety; monitoring methane and C:N ratio of input flow	Anaerobic digester, biogas storage equipment, and facultative and maturation lagoon(s)

3.9 ASSESSING REGULATORY ASPECTS OF SITE EVALUATION

Local governments are important in scaling up sanitation through the power of DWMS while protecting public health and the environment. While much of the actual work of site evaluation can be performed by the private sector or site owners themselves, local governments can establish enabling environments by eliminating roadblocks to compliance. This can be accomplished in resource-limited agencies by using a *self-monitoring regulatory* compliance system in which local governments provide the forms, instructions, educational resources, spot-checking, and oversight. However, this is complicated by how complex regulatory programs for DWMS usually already exist and can vary greatly from one country to the next, and even within countries.

Regulatory requirements also may have cost implications for the developer beyond the basic compliance requirements. Such requirements may include the following:

- pre-construction site investigation and inspection programs;
- permit review and collection of fees;
- inspections during and after construction; and
- operational inspections, sampling and analysis, and compliance audits.

Facility owners are generally responsible for ensuring that their systems are in compliance. This may require sampling, laboratory analysis, hiring certified operators, and completing compliance reports. Or there may be no requirements at all. Either way, it is important to understand the regulatory expectations to ensure compliance with all local requirements.

3.10 REVIEWING FINDINGS

This chapter provided information relating to:

- site planning and area availability estimation;
- soils depth, texture, structure and consistence assessment, as well as basic color analysis;
- groundwater and surface water as potential limiting conditions;
- slope and topography concepts;
- land use and utility/pre-existing DWMS concerns;
- access issues; and
- residual and regulatory considerations

Table 3.6 summarizes site evaluation findings related to the wastewater source (sample values from LORMA).

Table 3.6 LORMA Medical Facility site evaluation summary.

Aspects	Description	Local Condition
Available area	Determining how much land is available to help understand if a mechanized or non-mechanized system is best.	60 m^2; implies mechanized system is needed, especially given the high flow rate.
Soil properties	If soil dispersal is planned, determining the depth of usable soil and its key characteristics to calculate LTAR. If the system will be heavy, finding out if the soil structure is weak and needs additional support.	N/A – no soil dispersal is planned. Installation of concrete slab required to support weight.
Surface- and ground-water	Determining the impact on the site due to surface waters and flooding, since groundwater is not a major concern.	Flood elevation identified – system is elevated above it.
Slope and topography	Determining if gravity will enable non-pumped sewerage from point of generation to treatment site.	Force of gravity is insufficient; lift station is required.
Land use and utilities	Determining if the proposed DWMS is suitable based on local land use, zoning, and utility networks.	Phased approach acceptable to local authority and electrical utility access is convenient.
Access	Determining the need/potential for access for desludging, service, and inspection.	Site is fully accessible for all service providers.
Residuals and regulations	Determining if/how the DWMS managers could benefit from treatment residuals, and the need to take local regulations into consideration.	Agriculture- and biogas-based residual creation is not feasible, while local authority has given approval.

REFERENCES

County of San Diego Department of Environmental Health (2010). Design manual for onsite wastewater treatment systems. http://www.sdcounty.ca.gov/deh/water/docs/lu_Design_Manual_for_OWTS_3-22-10.pdf (accessed 10 Feb 2013).

Geospatial Training and Analysis Cooperative (2008). Calculating a slope. http://geology.isu.edu/geostac/Field_Exercise/topomaps/slope_calc.htm (accessed 25 June 2011).

Illinois Soil Classifier Association (2010). Understanding Soil. http://www.illinoissoils.org/Links_Files/Understanding%20Soils_Final2.pdf (accessed 20 November 2013).

Lindbo D. L., Stolt M., Miles R., Mokma D., Greene S. and Hoover M. (2005). Soil and Site Evaluation Module Text. http://www.onsiteconsortium.org/ed_curriculum/Practitioner/B/Soil_and_Site_Evaluation_Chapters_1-5.pdf (accessed 12 June 2013).

Lindbo D. (2009). *Chapter 4: Soil and Site Concepts for Installers. Installation of Wastewater Treatment Systems*. Consortium of Institutes for Decentralized Wastewater Treatment (CIDWT), Iowa State University, Midwest Plan Service, Ames, IA, 43–62.

Roth USA (2007). Buoyancy Calculations. http://www.roth-usa.com/PDF_Download_Files/RMT_buoyancy_calc.pdf (accessed 12 June 2013).

Chapter 4

DWMS technologies

This chapter will

- Explore the component categories of DWMS: user interface/pretreatment, sewers, primary treatment, secondary treatment, potential tertiary treatment, and the final disposition of the treated effluent through dispersal, discharge, or reuse;
- Introduce design and sizing strategies for key technologies;
- Provide guidance on determining the required level of treatment for specific needs.

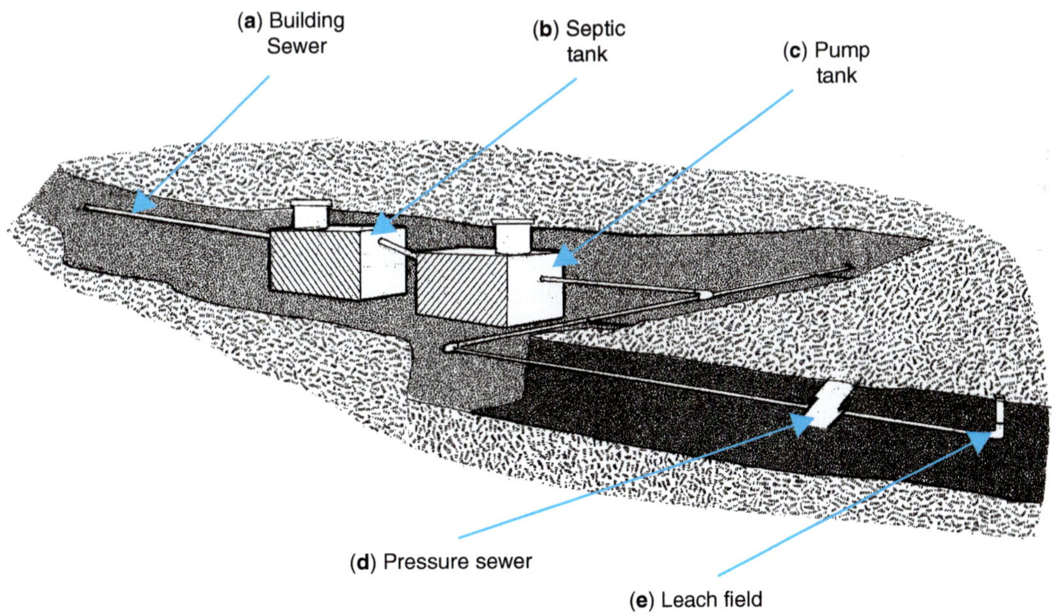

Figure 4.1 A typical DWMS illustrating how various components come together to form a system. In this case (a) the building sewer connects the building to the septic tank, (b) the septic tank provides primary treatment, (c) the pump tank pressurizes the effluent for distribution, (d) the pressure sewer conveys the effluent from the point of treatment to the point of discharge, and (e) the leach trenches provide secondary treatment and dispersal (Hoover, 2013, Lindbo 2013).

About Figure 4.1

Figure 4.1 is a reminder that wastewater management systems are really assemblages of different components that work together to achieve the project's goals. Defining these goals at the outset of the planning process can inform the technology decision making, as presented in Chapter 3. If the goal is to treat the wastewater to a level where it can be safely dispersed into the environment, technologies can be selected based on the source characteristics and site conditions to achieve that goal. If, however, the goal is to recycle the treated effluent for agricultural irrigation, collect biosolids for biogas production, or use them as fertilizer, other technologies may be more appropriate.

This chapter presents the major DWMS component categories as well as some of the technology options to provide the reader with an overall understanding of how systems are formed by various components that are specifically chosen to achieve desired results. Case studies are provided to illustrate how certain technologies have been previously used, and sizing math is given for a few of the technologies. It is not intended, however, as a comprehensive list of all the technologies, as that would be a very long list and is not what this manual is about. This manual intends to show how the source characteristics and site conditions influence technology selection, and we try to accomplish this through the introduction of carefully selected representative technologies.

Perhaps the overarching message of this chapter (and of the manual itself) is that proper DWMS practice is not about taking a favorite technology and adapting it to a given site. Instead it is about understanding the nature and goals of the project, the source and site conditions, the skill level of local service providers, and the governance realities of the host community, before selecting the most appropriate technologies based on those conditions.

Also, it must be said that salespeople can be aggressive in promoting their products. Purchasers and the hired service providers must become knowledgeable about the technology selection process to ensure that funds are spent appropriately and that the project's long-term goals are realized and sustained.

4.1 DWMS COMPONENT CATEGORIES

DWMS use various components, which may include configurations of pipes, tanks, basins, mechanical equipment, and other products. These components are selected based on a careful consideration of the source characteristics (Chapter 2), the site constraints (Chapter 3), and the needs of the end-user. DWMS are generally composed of four to five categories of sub-systems listed below:

- *User interface* – The toilet, which is the first step in the wastewater collection process for residential and other sources. There are many excellent references for toilet and latrine design, some of which are listed in the references at the end of this chapter.
- *Pretreatment sub-system* – Components that remove materials that are harmful to the treatment process at the source. Pretreatment devices are needed at many commercial facilities and include grease traps for food service, trash traps for markets, and lint traps for commercial laundry facilities.
- *Conveyance sub-system* – The network of pipes, channels, or sewers that transport the wastewater from the point of generation to the point of treatment, from one treatment component to the next, and then to the location of effluent discharge, dispersal or reuse.
- *Treatment sub-system* – A series of components that receive the wastewater and work together to break down and remove its contaminants, using a combination of physical, chemical, and biological processes. As shown below, this is a broad category, which includes the following:
    - primary treatment, which usually involves physical clarification by allowing solids to settle in tanks;
    - secondary treatment, which often uses microbes to consume pollutants through respiration; and
    - tertiary filtration and disinfection, required in some cases to make the treated effluent suitable for reuse or in some cases discharge or dispersal to the environment.
- *Dispersal, reuse, or discharge processes* – achieve the final disposition of the treated effluent, which may include the following:
    - dispersal of effluent through subsurface leaching beds or trenches, often considered themselves as secondary treatment;
    - reuse for a variety of beneficial activities, such as toilet flushing or irrigation for agricultural crops or landscape plants; and
    - discharge to off-site drainages, lakes, rivers or oceans.

4.2 TYPICAL DWMS COMPONENT CATEGORIES

Numerous combinations of technologies can form good site-specific DWMS. To better illustrate the concept of DWMS as integrated systems, consider the following illustration. Here a typical DWMS for a school is chosen to highlight how components are integrated (Figure 4.2):

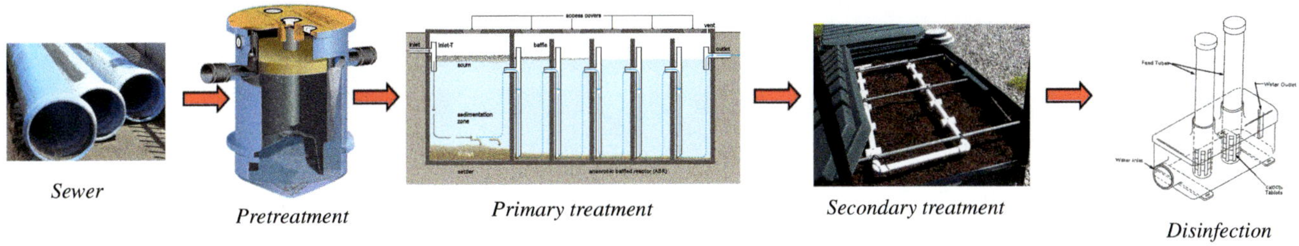

Figure 4.2 Wastewater system for a typical school, including (from left to right), sewer line, grease interceptor, ABR for primary treatment, biofilter for secondary treatment, and chlorinator for disinfection (Compendium, 2012; Thermaco, Inc., 2013; Authors).

4.3 USER INTERFACE AND PRETREATMENT

Decisions about which DWMS technologies might be best for a specific project begin at the source. This can start by considering the many toilet types, styles and manufacturers, as well as different models for how waste is collected and managed at the source. Ecological Sanitation (EcoSan) and certain composting toilets for example manage the wastewater right at the source. These concepts may be especially desirable for rural agricultural settings where the nutrients can be directly reused on crops.

More commonly, sewage management uses toilets to collect wastewater and send it through sewers to treatment systems. If source wastewater has pollutants such as high levels of grease, they should be removed by pretreatment devices located at the source to protect downstream equipment and capture these materials so that they may be reused or recycled. The sections provided below will introduce some of these technologies and concepts to illustrate the different methods of capturing and processing wastewater. Additional information is provided in the reference section or through links embedded within the text.

4.3.1 User Interface

For residential sewage, latrines have been the basic model of sanitation for centuries. However, given current focus on resource sustainability, toilets and DWMS have been the focus of much innovation in recent decades. Such innovation has aimed at improving cost-effectiveness and efficiency, particularly in the case of waterless toilet technologies and nutrient-capture programs such as EcoSan. While standard toilets can certainly serve as effective interfaces for DWMS, this discussion will focus on two models of toilet systems as a brief introduction to innovation in this area.

4.3.1.1 EcoSan Toilets

EcoSan treats human waste as a *resource* rather than *waste*, which is shown in how such toilets use a urine/feces separation and urine diversion design to enable the reuse of the settled and naturally-degraded wastes' nutrients for agricultural purposes. EcoSan projects have been implemented all over the world, mainly on a small scale, with both successes and failures. Low-income farmers in rural areas often benefit from using treated human waste as a free fertilizer source, so EcoSan projects have mainly been undertaken in these rural areas.

The urine and feces separation, which prevents contamination of the urine, occurs most commonly through the use of multiple-chamber urine-diverting toilets (Figure 4.3a). These have one hole or chamber to capture urine and another hole or chamber to either capture feces or to flush it away like in a regular toilet. Some of the toilets also add another hole for diversion of anal-cleansing water.

For most units, the feces are stored in a bucket located underneath the structure, while the urine is stored below the toilet in a sealed container (Figure 4.3b), where the urine can self-sterilize within 1 month. Afterwards, it can be diluted with water and directly applied to agricultural or ornamental crops as a nitrogen-rich fertilizer. The feces can be flushed to an existing DWMS, or for an improved approach, can be captured in a separate chamber (often lined with a plastic bag) and disposed of, composted, or dehydrated.

Figure 4.3 (a) Double-chambered EcoSan toilet (Doczi, 2010); (b) Urine-diversion mechanism (Doczi, 2010).

For a basic overview of urine-diverting toilets and other similar units, refer to the Compendium, pp. 42–73. For a more-detailed description of Eco-San, refer to Winblad & Simpson-Hebert (2004): https://wiki.umn.edu/pub/EWB/Uganda/SIDAGuidebook.pdf

4.3.1.2 Other Innovative Systems

In contrast to the approach of using the toilet without treatment to harness resources, several organizations are implementing or sponsoring systems that both treat the waste and capture resources from the treatment process. The most obvious example within the field is the Bill & Melinda Gates Foundation's Reinvent the Toilet Challenge, which is financing a handful of teams through grants of millions of dollars to make cost-effective, full-treatment, next-generation toilets.

One example of an innovative system is the LooWatt toilet block (Figure 4.4) that serves urban as well as rural areas. The LooWatt model equips community toilet blocks with composters and biodigesters, which combine human waste with wastes from food service facilities to maximize methane production that is used for cooking or heating. The biodigestion also yields biosolids that are composted with other waste streams. That leads to fertilizer products that provide extra value to the community. The LooWatt system shows how incentives that drive demand for sanitation services can make systems feasible in even the most land-constrained and difficult sites while improving health and economic opportunities. Like with other DWMS technologies, their implementation also requires knowledge of the wastewater source and the site where systems will be installed to ensure proper installation and byproduct usage.

Figure 4.4 Biodigester that treats toilet and food waste through the process of co-digestion (V. Gardiner, personal communication, June 11, 2013).

Find more information about the LooWatt toilet block system at http://www.loowatt.com/.

4.3.2 Pretreatment

Pretreatment removes certain pollutants from the wastewater to reduce its strength while protecting downstream components from damage or clogging. Common categories of pretreatment devices include

- grease traps and interceptors for food preparation, utensil washing, and food processing operations;
- lint traps for commercial laundry operations;
- trash traps for public markets or other facilities that may discharge a high concentration of suspended solids in the wastewater; and
- other devices for commercial applications to address industry-specific contaminants.

Pretreatment devices must be sized according to the design flow and the concentration of the contaminant loading, and be selected based on operation and maintenance (O&M) considerations, unit cost, ease of installation, and desire for recycling captured materials.

4.3.2.1 Grease Traps for Food Service

Grease traps and interceptors remove the fats, oil, and grease (FOG) that are byproducts of most food operations. FOG management is a major area of concern for restaurants and large kitchens, which if ignored can easily clog building sewers leading to the treatment systems and add to wastewater loads on downstream treatment works. There are three types of common grease management devices:

- under-the-sink (UTS) style grease trap;
- outdoor grease interceptor; and
- high-efficiency grease interceptor (Figure 4.5).

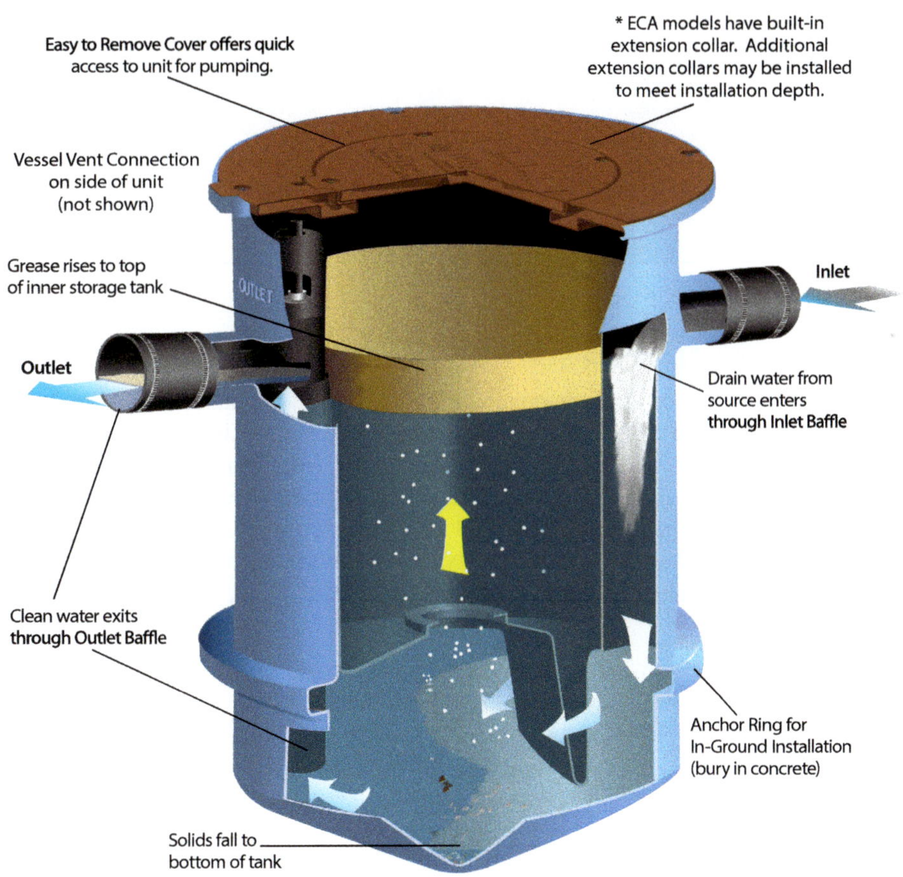

Figure 4.5 Trapzilla™ high-efficiency grease interceptor (Source: Thermaco Inc., 2013).

The capital and O&M costs, space required for setup, efficiency, and other aspects of these devices can vary significantly. Regardless of the type of unit, all require proper sizing, regular cleaning, grease collection and storage procedures, and, if possible, a procedure for recycling and processing the grease into value-added products.

For more technical and in-depth information, along with a grease management decision toolkit, refer to http://www.iwawaterwiki.org/xwiki/bin/view/Articles/WastewaterSolutionsDesignToolsforEngineers.

Grease management in Muangklang Municipality, Thailand

In 2011, the Municipality of Muangklang in northern Thailand instituted a commercial grease management program. It was a cooperative effort between the local government and commercial food service operators. The commercial operators installed grease traps, and the local government collected and dried the grease, which is used to power the community slaughterhouse, as shown in Figure 4.6. This is an example of a waste-to-energy program that created jobs and economic development opportunities and addressed a major pollution source in the municipality.

Figure 4.6 At left, grease trap before cleaning; at right, grease used to fire boiling pots at a slaughterhouse (Municipality of Muangklang staff, personal communication, October 12, 2012).

4.3.2.2 Lint Traps for Commercial Laundries

Commercial laundry facilities, as well as hospitals, resorts, and hotels with on-site laundry washing, should use lint traps to prevent fabric fibers and lint particles from clogging sewer lines and septic tanks. Lint traps should either be attached to the laundry machines or placed in concrete traps installed to collect the wastewater from all of the machines prior to discharge into the larger building sewer. Collected lint can be added to compost to improve soil for gardening, or if composting programs are not available, it may be collected and disposed of as solid waste. Additionally, commercial laundries that include dry cleaning must also separate their dry cleaning solvent wastewater from the building sewer, as that effluent is hazardous and can kill the bacteria that make biological wastewater processes work.

For more technical and in-depth information, along with best practice guidelines for wastewater-related lint management for commercial laundries, refer to http://www.iwawaterwiki.org/xwiki/bin/view/Articles/BestPracticesforCommercialLaundries.

4.3.2.3 Trash Traps for Public Markets

Public markets and other sources where high concentrations of suspended solids are present in the wastewater flow require trash traps to prevent solid waste from flowing down the drains and causing clogging problems. Traps must be cleaned frequently, while the system screenings should be placed in covered receptacles and disposed of as solid waste.

The main types of traps for markets are constructed with concrete and mesh, forming a box that strains and stores collected solids. This low-cost option can be made using locally-available materials and labor. Traps can be customized for different industrial uses to address specific wastewater pretreatment requirements. Owners of industrial facilities should check with industry trade groups for specifications on appropriate pretreatment devices like such traps and strategies for O&M.

Grease management at Red Ribbon Bake Shops in Manila, The Philippines

Red Ribbon Bake Shops is a chain of restaurants operating in the Philippines that use UTS grease traps in their facilities, which are full-service kitchens with food preparation and utensil washing. A typical facility's trap is plumbed for easy access, which is done by sliding out the work table (far left of Figure 4.7a). The grease trap is the metal box directly under the red bucket.

Figure 4.7 (a) Under-the-sink grease trap (Authors); (b) septic tank located underneath driveway (Authors); (c) grease trap cleaning time log sheet (Authors).

The restaurants incorporate grease management into their overall wastewater management strategy, which centers on a septic tank. It is located underneath the driveway (Figure 4.7b) and channels effluent to the common sewer system. Workers are required to check and clean grease traps once per day. Grease traps are manually cleaned, and the grease is placed in sealed plastic bags and handled as solid waste. The management follows a series of steps to ensure proper operation of the grease traps which include the steps given below.

(1) Managers provide training for all kitchen staff in proper grease trap cleaning practices. They also observe and log the time and date of the cleaning operations (Figure 4.7c).
(2) Workers dry-wipe greasy pots and pans prior to placing them in the dishwasher to minimize the amount of grease going down the drain.
(3) Management uses temperature control for operating the dishwashing equipment. Water temperature is kept under 60°C to help maximize the effectiveness of the grease trap.

For more technical and in-depth information, along with best practice guidelines for grease management in food service facilities, refer to http://www.iwawaterwiki.org/xwiki/bin/view/Articles/BestManagementPractices GreaseManagementforRestaurantsandCommercialKitchens.

General management guidance for pretreatment devices

- Ensure that the pretreatment device is sized for peak flows.
- Develop procedures for monitoring and servicing pretreatment devices.
- Determine in advance how wastes from pretreatment devices will be handled and disposed of.
- Determine if there may be beneficial uses for waste from pretreatment devices.
- Follow best management practices for grease traps and lint traps.

4.4 CONVEYANCE

Conveyance describes the transport of wastewater from the building to the treatment components, from one treatment component to the next, and then to final dispersal, discharge, or reuse. It can be achieved through human- and motor-powered emptying and carriage, typical of fecal sludge management programs; through combined pipes that carry both sewage and storm water; or through sanitary sewer systems that are separated from storm water flows. Only the network of pipes, channels, and associated equipment, rather than manual transport of fecal sludge, will be considered for the purposes of this manual.

Several types of sewer systems are commonly used within DWMS. Four applicable types are described in the following pages:

- *Building sewers* – The pipes and fittings connected directly to the buildings that transport sewage to other sewers or treatment works. Building sewers are also known as sewer laterals (Figure 4.8).
- *Simplified sewers* – Also called condominial sewers, they refer to a network of shared and relatively flat and shallow sewer lines that are laid often within property boundaries and not under roadways.
- *Solids-free sewers* – Otherwise known as settled, small-diameter, or effluent sewers, they use septic tanks for solids separation and transport the settled sewage to treatment works.
- *Pressure sewers* – Septic tank effluent pumping (STEP) systems, which use septic tanks and small pumping stations to collect, settle, and move the settled sewerage to treatment works through small diameter pipes.

Figure 4.8 Building sewers that lack cleanouts and sanitary sweeps, complicating their O&M (Doczi, 2012).

These are some of the more common types of sewers that may be available for specific DWMS. While large diameter (centralized) sewers for municipalities or dense urban centers may be either sanitary (wastewater only) or combined (wastewater and storm water combined), for DWMS applications, sewers are almost always designed for wastewater only.

4.4.1 Building and Gravity Sewers

As the most basic type of sewer, a building sewer (most often using gravity as the force mechanism) takes wastewater from the source to the first step of the site's DWMS. The overall strategy for building and other sewer design is to provide conveyance of the sewage from the point of generation to the point of treatment with minimal clogging. This is achieved by ensuring proper slopes for pipes, sizing the pipes to accommodate peak flows, using sweeps instead of bends, and providing enough cleanouts to perform routine and emergency maintenance.

For more overview of conventional gravity sewers, refer to the Compendium, pp. 94–95.

Basic guidance and a step-by-step process for installing these (and the other types of) sewers is given below.

(1) *Layout* – Determine the preferred route for the building sewer that will run from the building drain to the treatment system or common sewer. Measure the total length of the sewer run.

(2) *Slope* – The sewer line slope is typically 1% to 2% for pipes of around 20 cm in diameter, but this depends upon the size of the pipe. Other slopes may be considered on a case by case basis with proper engineering support. Determine the elevation of the flow line of the building sewer where it exits the building and calculate the distance from the building to the first component of the DWMS. This will set the elevation for the inlet to the DWMS.

(3) Conduct measurements as follows:
- stretch a level line from the building at the point where the sewer will leave it to the point where the sewer will connect with the sewer main or septic tank (Figure 4.9);
- attach this line to a stake at both ends to hold it above ground, and use a level to achieve a horizontal position;
- lower the end of the line at the main or septic tank the necessary distance to obtain the proper slope. For a horizontal run of 100 m and 2% typical slope, lower the line 2 m; for a run of 50 m, lower it 1 m; for 25 m, lower it 0.5 m, and so on (Figure 4.10); and
- cut a measuring stick to the length of the distance from the level line to the bottom of the drain where it leaves the building, and use this to guide the depth of the trench from the level line throughout its course.

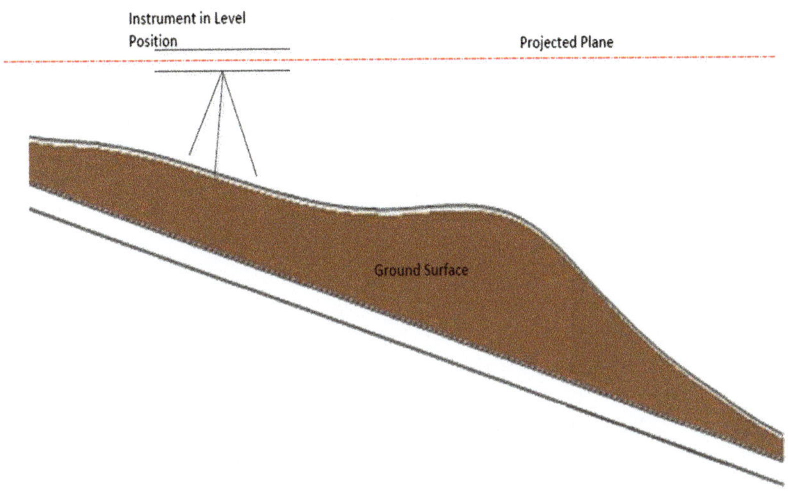

Figure 4.9 Using a building level to establish the horizontal plain.

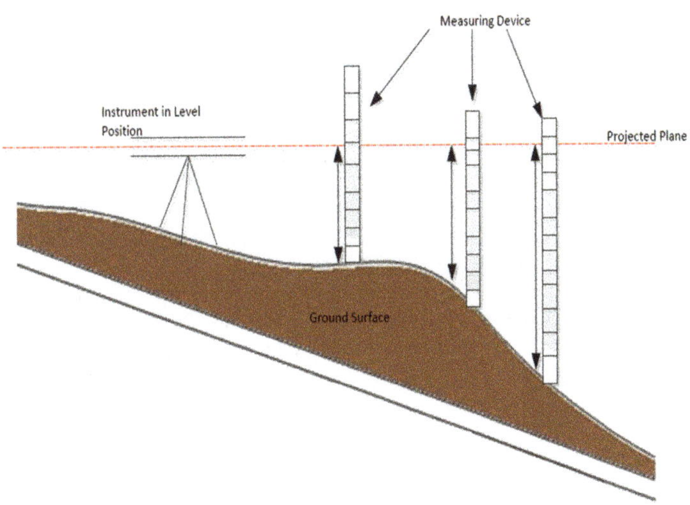

Figure 4.10 Using measuring devices to determine the elevation drop or rise from the horizontal plain at various points along the path of the sewer line, useful also for checking the slope of sewer line excavation beds (Durand, 2013).

(4) *Excavate the trench* – Be sure to over-excavate by 10 cm to allow for the bedding material.

(5) *Place the bedding material* – Use sand or pebbles and rake them to provide an even surface at the proper slope as described above.

(6) *Lay out the pipe in the trench* – Begin by attaching pipe, starting from the building. Be sure to install the cleanouts as needed, and take care to keep dirt and debris out of the sewer pipe.

(7) *Make the final pipe connection* – Check the slope again and adjust as needed by adding or removing bedding material from around the pipe. Make sure that the pipe joints are adequately supported with bedding material. When the proper slope is obtained, fill in any gaps underneath the pipe with bedding material.
(8) *Test the pipe for leaks* – Install caps on all cleanouts except the uppermost one, and plug the end of the sewer pipe. Fill the sewer line with water from the uppermost cleanout. Watch the water level in the cleanout pipe; if it drops more than 0.5 cm in 1 hour, repair the leaks. Since the trench has not yet been backfilled, leaks will be easy to spot.
(9) *Backfill the trench* – When satisfied that the pipe is laid at the proper pitch and is watertight, carefully fill around the pipe with remaining bedding material and cover with topsoil. Compact the soil to a slope that will allow surface water to flow away from the system.

Design guidance for DWMS sewers

- Design and plan sewer systems using locally-available sewer pipe materials.
- Design sewer systems with proper cleanouts – one within 1 m of each building, at every 15 m, and at lateral junctions, and 90-degree sweeps.
- Make sure all sewer line trenches are bedded with sand, aggregate base or compacted earth that is free of sharp objects that can damage pipes.
- Ensure that all sewer joints are properly connected (glued or pressure gasket-fitted) as per manufacturer's requirements and leak tested (described above) prior to backfilling.
- Backfill sewer trenches after the leak test is completed, and then compact and grade the trench soil (following the land contour) to drain surface water away from the sewer run.
- Account for how sewer type selection depends greatly on slope. Sewer lines that must run uphill will require pumps, while overly steep sites will require sewers to be properly stepped.

4.4.2 Simplified Sewers

Simplified sewers collect all household wastewater and convey it through gently sloping small diameter pipelines installed in front or back yards of houses. They are less expensive to install and operate due to how they incorporate:

- smaller diameter pipe (on the order of 10 cm diameter), which for long runs can result in significant savings up 20 percent of typical sewerage as shown in Bolivia (Foster, 2001);
- shallower depths and slopes (around 0.5%) that require less excavation and therefore result in cost savings of 45 to 75 percent relative to typical sewer excavation (Foster, 2001); and
- sewer line placement under property lines rather than under roads, which allows for increased household connections and less pipe usage.

Overall, these cost savings can reach 40 percent, even when accounting for the value of the community members' labor and O&M time, and water operator training of community members. This can result in a favorable cost-benefit ratios (Foster, 2001). Also, despite their small diameter and shallow slopes, these sewers can take quite large flows – even up to 120 m^3/d for 10 cm diameter pipes with 0.5% slope (Mara *et al.* 2010). However, users usually need additional education on the proper use of these sanitary systems, so that large solids, trash, and grease are kept out of the wastewater.

Simplified sewer programs rely upon well-coordinated and effective resident-led O&M programs, as the systems are intended to be customized for and managed by the communities themselves. This mainly involves clearing out blockages and checking system functioning by accessing the inspection chambers, which serve as the replacement for the costlier manholes used in conventional gravity sewers. Typical programs assign responsibilities to local workers who maintain the shared sewer laterals in exchange for reduced sewer bills. City or utility officials maintain responsibility over the O&M of the larger sewer mains.

For more overview of simplified sewers, refer to the Compendium, pp. 90–91.

4.4.3 Solids-free Sewers

An extension of the simplified sewer concept, solids-free sewers have small-diameter (at least 7.5 cm), low-slope piping and rely on gravity and community O&M (Compendium, 2014). However, they require a septic tank or other primary treatment device to better remove influent solids. Solids-free sewers are relatively low cost and simple to design, install, and maintain and have greater flexibility than simplified sewers because they have a lower risk of clogging. This permits a design that can use shallower and more variable slope. Since septic tanks are needed, these systems require organized desludging programs.

For a basic overview of solids-free sewers, refer to the Compendium, pp. 92–93. For more technical and in-depth information refer to Alternative Sewers: A Good Option for Many Communities (Casey, 1996) pp. 5: http://www.nesc.wvu.edu/pdf/WW/publications/pipline/PL_FA96.pdf.

4.4.4 Pressure Sewers

The traditional view of development experts has been that passive sanitation systems are better than mechanized technologies, and that for sustainability reasons, technologies that require electricity or have moving parts should be avoided. As a community's technical capabilities and capacity increase, it can look to new and better ways of solving wastewater problems. In some instances, using simple effluent pumping stations can enable these alternative approaches, which in some instances can save money and actually enhance sustainability when compared with non-mechanized systems. This is especially true in space-constrained areas where using an effluent pump makes it possible to provide more complete wastewater treatment at better locations or to use certain low-cost technologies that require less space.

In general, there are two types of mechanized (pressure sewers). Those that

- use grinder pumps to chop up the raw sewage into fine particles and then pump the slurry through small diameter pipes, otherwise known as grinder systems; or
- settle out the solids first through properly designed septic or settling tanks, similar to solids-free sewers. Since these also use pumps, they are known as septic tank effluent pumping (STEP) sewers.

Grinder sewer systems are more expensive than STEP sewers and require more frequent and specialized O&M. For these reasons, they are not advisable for most developing country settings. When pressurized systems are required, STEP systems (Figure 4.11) are recommended, because among other benefits, they use small diameter pipes, between 4 to 7 cm, much lower than the typical 20-cm minimum for conventional gravity sewers (Casey, 1996). In general, when designed and installed properly, STEP systems can be highly reliable, lasting many years before parts require replacement. When local service providers are trained in the proper installation and maintenance, STEP systems become part of the service delivery and local sanitation value chain that can help drive market-based sanitation improvement on a wide scale.

Figure 4.11 Community STEP sewer system schematic (Jones *et al.* 2001).

For a summary of STEP sewer system technology and practice, refer to Alternative Sewers: A Good Option for Many Communities (Casey 1996) pp. 3–4 at http://www.nesc.wvu.edu/pdf/WW/publications/pipline/PL_FA96.pdf.
For a detailed overview of, design guidance for, and information about pressure sewers, refer to the US EPA (2002a) Wastewater Technology Fact Sheet, Sewers, Pressure: http://water.epa.gov/scitech/wastetech/upload/2002_10_15_mtb_presewer.pdf.

4.4.4.1 STEP Sewer System Components

The following sections describe the components that make up STEP systems. The components include:

- pump vault to house the pump and other equipment;
- screens to protect the pump from clogging;
- a pump, typically a submersible effluent pump; and
- electrical components including the pump controller, high water alarm, junction box, wiring, conduit, and control panel.

4.4.4.1.1 Screened Pump Vault

The pump vault is the housing for the pump, which is the center of the STEP sewer system. A pump vault may have many configurations. One configuration (Figure 4.12) has an outer structure that houses the pump and the floats.

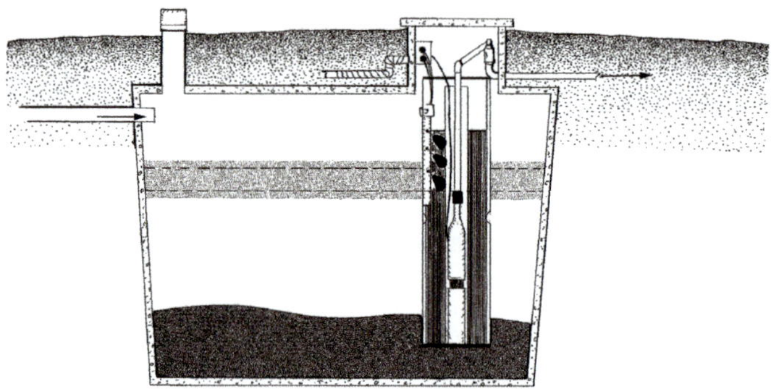

Figure 4.12 Screened pump vault complete with floats and electrical connection (Hoover, 2013).

Screens are maintenance items that require periodic cleaning. Typically screens should be cleaned by washing with a high-pressure hose every 3 months. More frequent cleaning may be required for larger flows, or if poorly designed or inappropriately sized septic tanks come before the pump system.

Design guidance for pump vaults

- If it is large enough, consider using the second compartment of the septic tank as the pump vault. This will eliminate the need for a second tank.
- Determine if flow equalization is beneficial to the overall goals of the wastewater system. If so, ensure that the pump vault is sized to hold the required volume to achieve flow equalization.
- Pump vaults, like septic tanks, must be installed on a compacted, level foundation and backfilled with sand or earth that is free of rocks and debris. This helps prevent differential settling or breakage.
- Pump vaults must be leak-tested prior to being placed into service and protected from buoyancy forces in high groundwater areas.
- Take extra care to ensure against infiltration from storm water by making all fittings watertight and diverting surface water away from the tank.
- Vent the vault to the building sewer only if it can be connected to a properly plumbed roof vent stack. Otherwise, install a dedicated vent that discharges septic gases at roof level. Avoid odor problems by never venting pump vaults at street level.
- Ensure easy access through gas-tight locking lids.
- Screens should be constructed out of materials that will not corrode or degrade in the wastewater environment. PVC or other plastic materials are best.
- Choose a screen mesh that will allow water to pass freely but eliminate the larger solids; 1–2 mm mesh opening is recommended.
- Install pump and alarm float switches to ensure that minimum volume, operating volume, and reserve volume requirements are met (Figure 4.13).

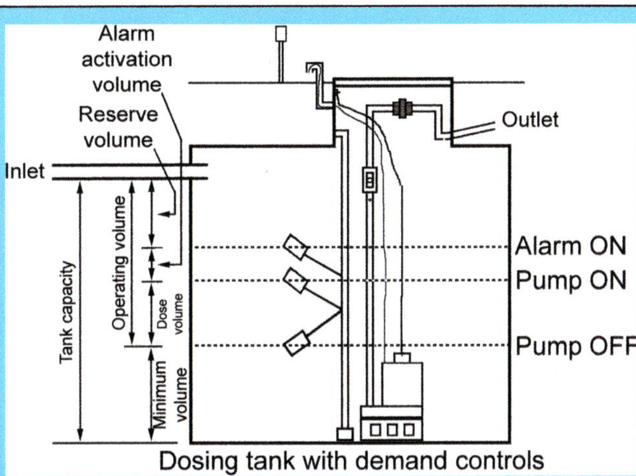

Figure 4.13 Components of a dosing tank (Consortium of Institutes for Decentralized Wastewater Treatment, 2009).

4.4.4.1.2 Pumps

The pump can come in various models with different electrical requirements. Sump pumps and turbine pumps are two of the more common pump types used in pressurized sewer systems:

- Sump pump: The submersible-style sump pump is the most common type for pressurized sewers. The intake from the sump pump is at the bottom, and it is often placed on a concrete block so that it sits well above the level of any accumulated sludge. Caution must be taken to ensure that sump pumps are 'effluent rated'; however
- A better option may be a centrifugal high head effluent pump, fitted with Teflon™ seals to prevent corrosion.

Whichever pump is used, make sure that the float switches (three of which are shown in the middle of Figure 4.13) are set so that the pump is always fully submerged. This helps keep the pump cool and lengthens pump life.

Design guidance for pumps

- Select a pump that is adequately sized to deliver the flow rate desired given the elevation of the rise and length of the run of the sewer path. Verify that the pump is adequate to the task prior to purchasing.
- Be sure to match the pump with the voltage and phase of the available power source.
- There are several styles of pumps for low-pressure systems, including turbine pumps, displacement pumps, sump pumps, jet pumps, and others. Check to verify that the pump is rated for effluent. This typically means that they are constructed with Teflon™ seals and other components that do not corrode.

4.4.4.1.3 Electronics

4.4.4.1.3.1 Float switches. Float switches control the on-off operation of pumps and alarms. For pumps, the float switch is wired in-line between the controller and the pump. Two styles of float switches are commonly available. The first uses mercury to make the electronic connection. The second uses a metal ball built inside of the switch housing to make the connection. In addition, there are normally 'off' switches (where the electrical connection is broken when the float is in the down position), and normally 'on' float switches where there is a positive electrical connection when the float is in the down position. For pump controllers, use a normally off float switch. When the water rises in the tank, the float switch rises on its tether, and the angle causes either the mercury or ball to complete the connection, and the pump will operate, at least until the water level in the tank drops down and the switch turns off.

Float switches are available as narrow-angle and wide-angle switches. Figure 4.14 illustrates these two common switch configurations, each with the option for having the floats be tethered to an independent fixed object (float trees) or be attached to weights. Either way is approved for general use in STEP systems applications.

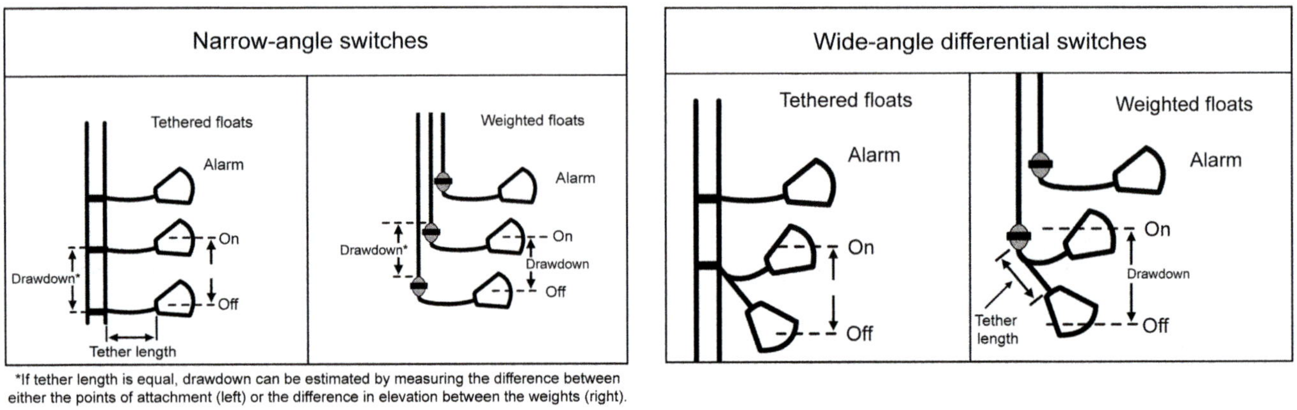

Figure 4.14 Float switch arrangement using weights or float trees with narrow- or wide-angle arrangement (Lindbo, 2013).

Design guidance for float switches

- Tether the float at an elevation to maximize storage volume in the tank.
- Verify that the off level is above top of the pump. This will ensure that effluent always covers the pump, which helps keep it from overheating and corroding.
- Set float switches to match the required dose and test.
- Use a redundant off float switch, which will protect the pump from getting dry in the case of unexpected low water conditions. This can be part of a series of switches, similar to the one shown in Figure 4.13.
- Secure pump floats using weights, which are easy to set and adjust, or a 'float tree' that can be removed independent of the pump.
- When possible, use mechanical floats to limit the potential for mercury entering the environment.

4.4.4.1.3.2 High water alarm. High water alarms are used to indicate abnormal operating conditions of the pump station, such as pump failure, pump switch failure, or excessive flow to the system. High water alarms use float switches similar to the pump control switch that, when activated, may send a visual alarm (flashing red light), audible alarm (siren or other sound), and/or a signal connected to a remote monitoring device that sends an alert to a Smartphone.

Design guidance for alarms

Locate the alarm float in the pump tank so that the 'alarm on' position is just above (within 15 cm of) the pump activation level. This allows for maximum emergency storage capacity in the tank and time for maintenance before sewage backs up into facility or overflows. The only exception is for use in an equalization panel, where additional space for effluent storage is designed into a system, and thus needs to be below the alarm level, while still providing sufficient emergency capacity above the high water alarm.

4.4.4.1.3.3 Junction boxes. Junction boxes are sealed boxes that are located outside of the tank and allow for watertight electrical connections. For pressurized sewer systems, the pump and pump control float are wired together in the junction box. Junction boxes are often made of plastic or PVC and come equipped with gas-tight sealed covers.

Design guidance for junction boxes

- Use plastic or PVC junction boxes to minimize corrosion.
- Use conduit seals or silicone caulk that will not shrink (non-shrink caulk) to seal the wire transfer ports to keep out moisture and sewer gas.
- Use waterproof wire nuts to make secure connections.
- Ensure that junction boxes are elevated above the high water mark for flood prone areas and protect them when possible from storm water.

4.4.4.1.3.4 Control panel and accessories. Controllers manage the pump on and off cycles from the input received from the float switches. They may be digital or analog (Figure 4.15), and may have a number of accessories, such as:

- elapsed time meters and cycle counters, which help operators keep track of flow and identify abnormal operating conditions;
- timers, which can be used to regulate the number of on-off cycles (wastewater dosings) per day and duration of each cycle. They are especially useful for flow equalization;
- event recorders, which identify a number of alarm conditions and pump on-off cycles; and
- remote monitoring mechanisms.

Figure 4.15 Control panel with visual alarm, cycle timer, and separate breakers for the pump and alarm (Authors).

Design guidance for control panels

- Like with junction boxes, the controllers should be protected from sewer gas by using conduit seals or non-shrink silicone caulk at all conduit connections.
- Separate circuit breakers for pumps and alarms should be used to ensure that the alarms will continue to function should the pump breaker trip.

4.4.4.1.3.5 Potential tank for flow equalization: technical considerations. Flow equalization, introduced in Chapter 2, uses a pump system configuration similar to what has been described for STEP systems (Figure 4.16). For

STEP system applications, the operating volume (the average daily volume plus the surge volume – the extra flow that needs to be equalized) sets the overall tank capacity requirement. While the same equipment discussed thus far is used, a specialized control panel with a programmable timer is also required.

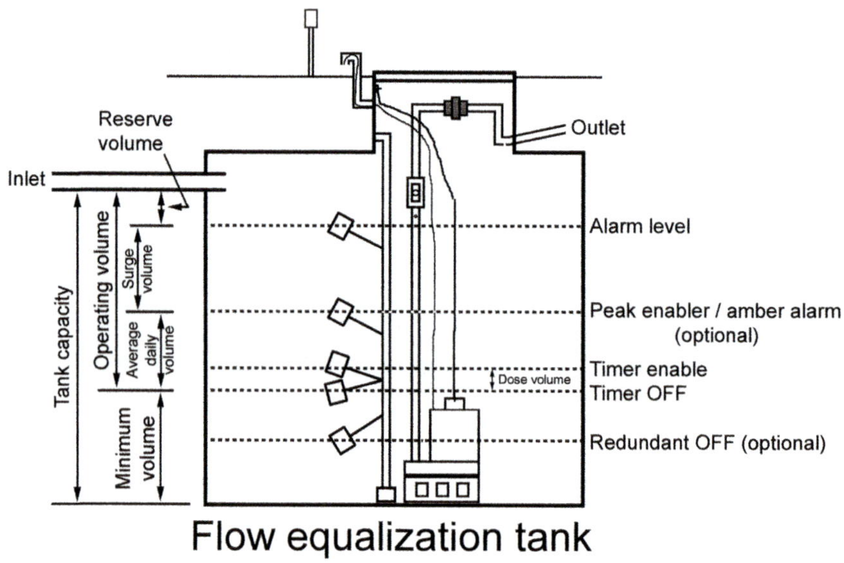

Figure 4.16 Float settings and considerations for a flow equalization tank (Consortium of Institutes for Decentralized Wastewater Treatment, 2009).

4.5 PRIMARY TREATMENT

Following pretreatment, primary treatment achieves the initial reduction of organic material and suspended solids so that those constituents do not limit the efficiency of later treatment steps. The major primary treatment technologies are septic tanks, anaerobic baffled reactors (ABRs), biodigester systems, and anaerobic lagoons. While those lagoons can independently achieve solids removal, they are mostly used in series with other types of lagoons in areas with ample land, which is why they will be further described in Section 4.6 (for secondary treatment).

4.5.1 Septic Tanks

Septic tanks are low-cost, watertight chambers made of fiberglass, concrete, PVC, or other plastic materials that reduce solids and organics through settling and anaerobic digestion. Septic tanks, an example of which is shown in Figure 4.17, have a minimum of two compartments separated by a baffle that helps retain solids and scum. The scum is mostly made up of cooking oil and grease, which floats to the top, while the heavier solids sink to the bottom to form a sludge layer.

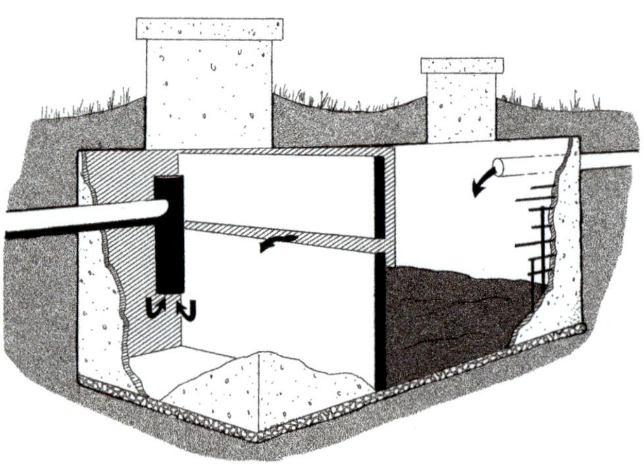

Figure 4.17 Septic tank cross-sectional view (Hoover, 2013).

Septic tanks encourage anaerobic digestion, which helps break down the solids over time. However, since the rate of solids accumulation is often greater than that of their digestion, the sludge must be emptied periodically. This implies the need for easy accessibility for emptying by vacuum trucks.

For more information on septic tanks, refer to the Compendium, pp. 74–75, while more detailed discussion can be found in Sasse (1998), pp. 69–72. For more technical information, refer to US EPA (2000a) Decentralized Systems Technology Factsheet: Septic System Tank: http://water.epa.gov/scitech/wastetech/upload/2002_06_28_mtb_septic_system_tank.pdf. An additional technical reference and in-depth information on these systems customized for high flows can be found in US EPA (2000b) Decentralized Systems Technology Factsheet: Septic Tank Systems for Large Flow Applications: http://water.epa.gov/scitech/wastetech/upload/2002_06_28_mtb_septic_tank_large_flow_app.pdf.

4.5.2 Anaerobic Baffled Reactors

ABRs can be considered larger (and improved) septic tanks. They use some of the same processes, but can achieve higher efficiency. ABRs, an example of which is shown in Figure 4.18, route the influent upwards through the granular sludge blanket, where solids are trapped and consumed by microbes. ABRs can be used for larger flows than septic tanks and generally produce higher-quality effluent. Like septic tanks, ABRs are relatively low cost, can be constructed using locally available materials and labor, and need to be desludged periodically.

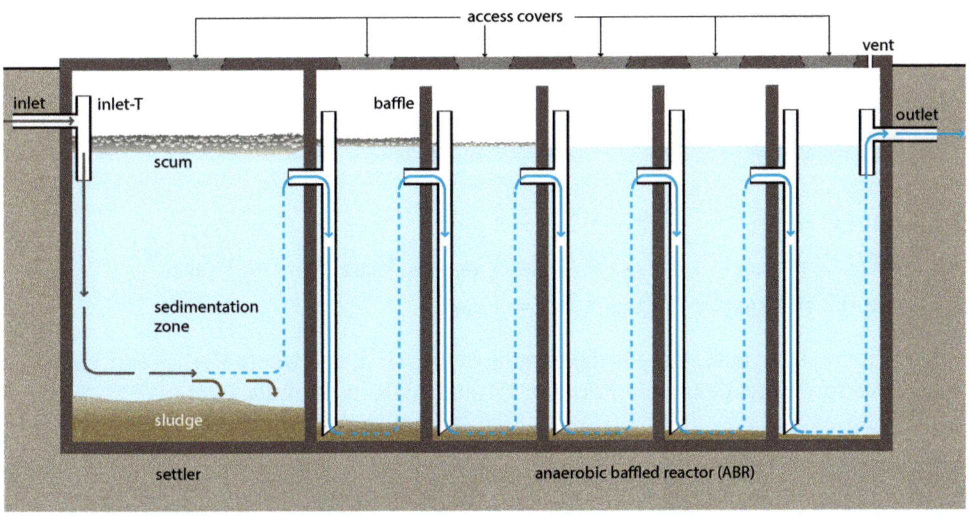

Figure 4.18 Typical ABR configuration (Compendium 2014).

Planning guidance for septic tanks and ABRs

- Size and install them using the guidance described below, taking care to ensure proper volume, water tightness, and anti-flotation measures.
- Ensure all compartments of the tank are accessible to desludging equipment. Locate the tank so that it can be reached by vacuum truck hoses or determine other desludging arrangements prior to installation.
- Vent them properly, either through the vent stack in the building plumbing (if present) or through dedicated vent stacks that vent sewer gas at roof level, rather than street level.
- Ensure that access ports or manholes are located over each compartment to enable desludging. Access ports should be gas-tight and equipped with locking lids.

For additional basic overview of ABRs, refer to the Compendium, pp. 76–77, while more detailed discussion can be found in Sasse (1998), pp. 79–82. A descriptive thesis on the relationship of septic tanks and ABRs, with examples of their application and related regulations in Thailand, written by Sri-Anant Wanasen (2003), is located at http://www.sswm.info/sites/default/files/reference_attachments/SRI%20ANANT%20W%202003%20Upgrading%20Conventional%20Septic%20Tanks%20by%20Integrating%20In%20Tank%20Baffles.pdf.

4.5.2.1 Design Strategy for Septic Tanks and ABRs

There are three major variables that dictate the design features of septic tanks and ABRs:

- flow (Q), which is the volumetric loading rate of the influent, given in Liters/d (L/d) or cubic meters per day (m^3/d);
- hydraulic retention time (HRT), which is the time in days a typical wastewater particle spends and gets treated within the septic tank or ABR; and
- desludging interval (DI), which is the time interval in years – assumed to be one for the purposes of conservative design – that the septic tank or ABR can operate without desludging, which is usually up until sludge reaches 30 percent of overall ST capacity.

Many related formulas drive the design of septic tanks and ABRs. Two common ones, shown below, suggest coefficients for compartment sizing. While two-thirds and one-third values for septic tanks are well accepted, the values for ABRs are just suggestions, as site-specific engineering is required. For the ABR Formula (1.2), the coefficient value for the first chamber, which can be thought of as the settling chamber, is double the other compartments' values, which indicates that it is twice as long as the others. In general, the length of the settling chamber is permitted to be double or triple the length of each of the other compartments, of which there need to be at least four in total for most scenarios.

Other commonly used guidelines are that the total length of the septic tanks and ABRs should generally be at least at least 2 times greater than their width, and that their height is between 1.5–2 m. Also, tank liquid capacity (LC) is taken to be 0.8 (equals 80 percent) for the purposes of this manual. It refers to the proportion of the septic tank or ABR that should be filled with liquid/sludge, with the rest being filled by air.

Formula 1.1: $Q * \mathrm{HRT}/\mathrm{LC} = V_{\mathrm{ST}}$; $V_{\mathrm{ST}} * 2/3 = V_{\mathrm{ST1}}$; $V_{\mathrm{ST}} * 1/3 = V_{\mathrm{ST2}}$;

Formula 1.2: $Q * \mathrm{HRT}/\mathrm{LC} = V_{\mathrm{ABR}}$; $V_{\mathrm{ABR}} * 0.4 = V_{\mathrm{ABR1}}$; $V_{\mathrm{ABR}} * 0.2 = V_{\mathrm{ABR2}}$; $V_{\mathrm{ABR}} * 0.2 = V_{\mathrm{ABR3}}$; $V_{\mathrm{ABR}} * 0.2 = V_{\mathrm{ABR4}}$;

(Where V_{ST} is the volume of the septic tank, V_{ABR} is the volume of the ABR, and where $V_{\mathrm{ST1...2}}$ and $V_{\mathrm{ABR1...4}}$ are the volume of the individual compartments for the septic tank in Formula 1.1 and ABR in Formula 1.2).

Septic tanks/ABRs and their relationship to soils and groundwater features

- *Soil depth:* Be sure to excavate test holes in the area where tanks will be installed to verify that the soil depth is sufficient to install the tanks. Rock outcroppings or bedrock in the areas of the proposed tanks can require special measures to achieve proper tank hole depths.
- *Groundwater:* Presence of groundwater, or the potential that seasonally high groundwater might impact tanks, requires special attention during design and installation. Sealing the outside of the tank to prevent infiltration and protecting the tank from flotation due to buoyancy forces is crucial.

4.5.3 Anaerobic Digesters

Biodigestion is an anaerobic fermentation process of wastewater treatment that generates methane-rich biogas as a byproduct. For DWMS applications, anaerobic digestion is mostly used for livestock operations and residences with livestock, as the process requires high-strength wastewater to generate usable volumes of methane. Small plot farmers throughout Asia and Africa have been using the process for managing their human and farm waste for decades with good results. The systems require continuous O&M, but the reward for properly running systems is usually at least enough energy to satisfy a family's cooking needs.

The single chamber, usually household-scale digester structure comes in three major models: floating drum, fixed dome, or plug-flow tubular system (Figure 4.19) shows the last two types. While they all have their own advantages, disadvantages, and regions of greatest usage, they all anaerobically degrade wastewater, manure and/or green (crop) waste through the process of fermentation, which forms methane that rises from the sludge to a gas collection system. This feature requires extra attention in the design process to ensure the safe collection of captured methane. Such systems are equipped with pressure release valves, and in larger units, flaring systems to protect workers and facilities from catastrophic explosions. In general, designing and implementing these features require advanced site-specific engineering from trained personnel.

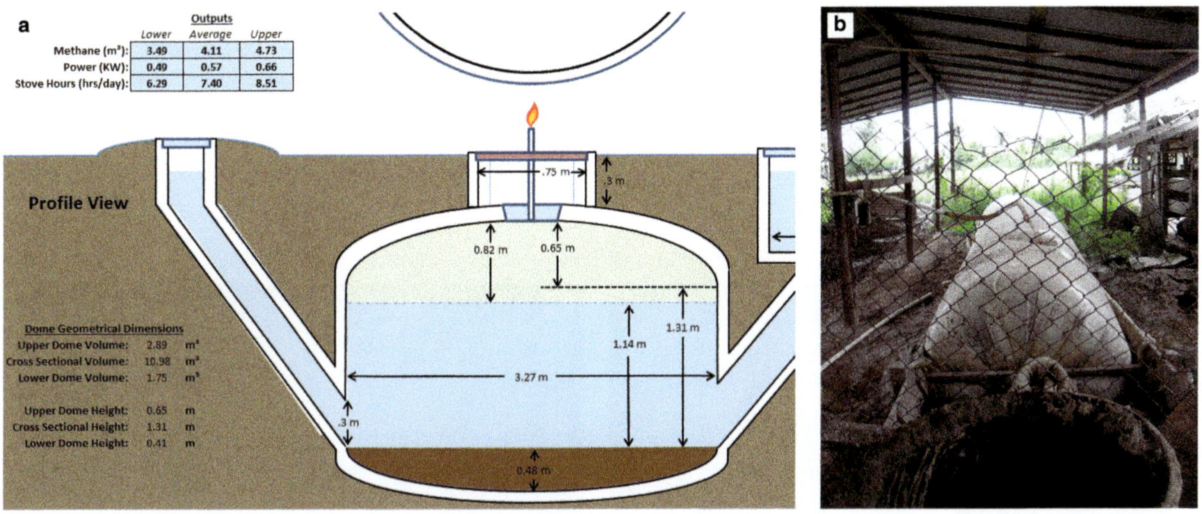

Figure 4.19 (a) Screenshot of Co-digestion Creator planning tool's horizontal view schematic of dome digester, showing estimated synergistic gas production values and necessary unit dimensions based on customized input waste scenario (Authors); (b) Plastic plug flow-type anaerobic digester with proper roof and fence protection in Guyana (Authors).

Along those lines, the units generally have greater construction costs than septic tanks and must be designed and installed by skilled engineers and trained contractors to ensure proper long-term function. This is even more the case for the larger-scale biodigester units, covered lagoons, and complete mix silo-shaped units that accept livestock waste from large farms.

While biodigestion focuses on a single feedstock, such as livestock manure, research into co-digestion, or the simultaneous biodigestion of multiple feedstocks in one reactor, shows promise. This process requires a balance between carbon, best obtained from green or crop wastes, and nitrogen, best obtained from animal waste. FOG is an additional waste feedstock, which can add a significant carbon boost to livestock waste and dramatically increase methane production. This is significant from a community-based DWMS program where organized FOG collection activities could support co-digestion programs as important energy sources for communities.

Digester sizing equations depend on the commonly used variables of flow rate and strength (based on the selection of the number or mass of animals, humans, and/or green or food wastes), desired hydraulic retention time, and temperature. The same is true for the sizing and structural design of these units that are beyond the scope of this text. Design guidance is given in the authors Co-digestion Creator, which is an (expanding) toolkit useful for sizing the Chinese Dome biodigester model and the plug-flow tubular system (which are both proven technologies for small flows). It allows for addition of both brown and green waste based on the C:N ratio to maximize methane generation. It can be accessed here: http://www.iwawaterwiki.org/xwiki/bin/view/Articles/RTIsCo-digesterCreatorV30beta

For more information on all aspects of the major household biodigester models and examples of their implementation and commercialization in Asian and other countries, refer to the SNV (a Dutch non-profit development organization) website: http://www.snvworld.org/en/sectors/renewable-energy/about-us/snv-energy/domestic-biogas/snv-and-domestic-biogas.

Co-digestion for increased methane production in Vietnam

Since the 1980s, researchers in the Mekong Delta region of Vietnam have been experimenting with co-digestion. This usually involves the digestion of wastewater with crop (green) waste and manure (brown) wastes in one reactor. The system shown here, developed at Can Tho University, is known as the V-A-C-B-D (farming, aquaculture, livestock, biogas, and treatment) method, in which wastes from pigs and humans are co-digested to produce methane-rich biogas for use in cooking, and nutrients that are recycled to help raise fish.

The anaerobic digester, shown here at the right (Dr. Chiem, the technology developer of Can Tho University, is in the foreground of Figure 4.20a), is a plastic tube, about 1 m in diameter and 8 to 10 m long depending on the size of the operation (blue and white tube shown partially buried). For this installation, the digester receives waste from 13 pigs and a family of 2 adults. The system is scalable to some degree, as larger families can use a larger reactor bag (or multiple ones in parallel), as long as the basic dimensions and principles of sound design and installation are adhered to. The vertical bags shown at the right side of the photo are optional gas storage bags. The biodigester costs approximately two million Vietnamese dong (around $100 US dollars), which will drop as the commercialization and mass-production of digesters accelerates there. This represents about half a month's income from backyard farming operations typical of the Mekong region.

Figure 4.20 (a) Plug-flow tubular digester in Vietnam (Authors); (b) Wetland trials at Can Tho Vietnam for secondary treatment for digester effluent (A. Thomson, personal communication, August 30, 2013).

In recent years, Dr. Chiem and his team have been developing secondary treatment for digester effluent. There are ongoing experiments with biofiltration and different configurations of constructed wetlands (Figure 4.20b) being tested at the facility. In the case of the wetlands, researchers from Can Tho University and Duke University (Ashley Thomson) in the United States are looking at nutrient and pathogen reduction for wastewater passed through horizontal planted beds constructed of gravel and coconut processing waste.

4.5.4 Other Primary Treatment Options

This section has focused on some of the more important technologies used for primary treatment in DWMS installations. There are other potentially appropriate technologies that also depend on settling and/or biodigestion, Two major examples are upflow anaerobic sludge blanket reactors (UASBs) and anaerobic filters (AFs).

UASBs depend on steady external water flow or electricity to drive a pump to force wastewater up through a suspended sludge blanket. The blanket contains bacteria that, based on expert-driven, sustained O&M, can degrade solids and inactivate some pathogens. AFs are like septic tanks, but their second chamber has a gravel filter hosting bacteria (called a fixed film system) that act like the bacteria in UASBs. Both UASBs and AFs have the potential to be much more effective in terms of solids and BOD removal than the previously-mentioned primary treatment technologies. Also, they combine the principles of primary treatment with some of the processes of secondary treatment discussed below.

For more information on UASBs and AFs, refer to the Compendium, pp. 122–123 and 108–109, respectively. Also, review the technical details of both of these technologies by visiting Sustainable Sanitation and Water Management at their comprehensive website. Explore both topics at: (Spuhler, 2008) http://www.sswm.info/category/implementation-tools/wastewater-treatment/hardware/semi-centralised-wastewater-treatments-7 and (Spuhler, 2010a) http://www.sswm.info/category/implementation-tools/wastewater-treatment/hardware/semi-centralised-wastewater-treatments/u.

Primary treatment technology selection and installation considerations

Determining which primary treatment technology to use is generally a function of flow rate, waste strength and the desirability of biogas as an end product. Small residential DWMS serving five homes or fewer will typically utilize septic tanks for primary treatment. Larger residential and commercial systems may use ABRs or other anaerobic technologies (especially anaerobic lagoons if land is readily available). If the waste composition (usually containing animal manure) is sufficient and methane generation desirable, one or more anaerobic digesters may be preferable. Whichever technology is used, removal efficiency of the organic matter and suspended solids will usually be between 40 percent and 70 percent, while no significant reduction in pathogens can be expected.

To properly set up the chosen primary treatment system, the user must:

- locate primary treatment devices in areas that are accessible to desludging equipment;
- include enough cleanouts (manholes) to accomplish desludging, relevant particularly for septic tanks and ABRs;
- ensure that those type of structures' access ports have gas-tight lids that are secure from unauthorized entry (using locks);
- use watertight construction practices. Seal the outside of tanks in high groundwater areas to prevent infiltration;
- provide a stable foundation on compacted soils where tanks will be installed to limit differential settling. Use concrete footers or foundations in high groundwater areas to secure against flotation;
- provide proper venting of primary treatment tanks. Ensure that there is a sewer gas pathway from each compartment of the tank to the vent. Use the plumbing stack vent in the home only if proper 'p' traps are provided for all plumbing fixtures that drain. Otherwise, install a dedicated vent terminating above the roof peak; and
- never vent sewer gas at street level; only roof level venting is approved.

4.6 SECONDARY TREATMENT

The moderate reductions in organic matter and solids achieved through primary treatment result in an effluent with pollutant levels that still exceed those allowed for safe discharge and reuse. Reducing those levels further requires secondary treatment, which are generally aerobic processes that utilize oxygen and mixing to stimulate aerobic bacteria to rapidly reduce organic matter as well as pathogen levels. Many different types of systems can accomplish this, such as soils-based systems, sewage lagoons, constructed wetlands, biofilters, and aerobic treatment systems that mechanically oxygenate the effluent.

The presentation of secondary treatment technologies below will describe the following:

- soils-based dispersal technologies, including leach trenches, similar alternatives, mounds and drip-effluent systems;
- constructed wetlands;
- sewage lagoons (waste stabilization ponds);
- biofilter systems; and
- aerobic treatment systems.

It is important to note that new technologies for wastewater treatment are always emerging, some of which appear to be very promising. New technologies may be considered for use in DWMS systems, but implementers should exercise caution and adopt them only when sufficient supporting data from actual functioning systems are presented.

4.6.1 Soils-based Dispersal Systems

Soils-based treatment systems send primary treatment (usually septic tank) effluent to leaching trenches or beds where it percolates through the soils and receives treatment. The treatment is achieved mainly through microbial degradation, as bacteria living in the soil are stimulated by oxygen that fills the pore spaces. If land area is plentiful, soils are deep and well-drained, and there are no foreseen beneficial reuse opportunities for the treated effluent, these may be the most effective and economical means of safe effluent dispersal.

4.6.1.1 Leach Trenches

Leach trenches (shown in Figure 4.21 in their typical series with a septic tank) are constructed with a series of pipes with perforations that evenly disburse the septic tank effluent within the trench. Effluent flows through the gravel or other medium and becomes absorbed in the unsaturated soils (vadose zone as labeled in Figure 4.22), which accomplishes the secondary treatment. However, if the soils are saturated with groundwater a clogging layer may rapidly form, and there will be little or no pathogen reduction, threatening environmental and public health. For this reason, extra care must be utilized to ensure soils are adequate for on-site leaching prior to installing these systems.

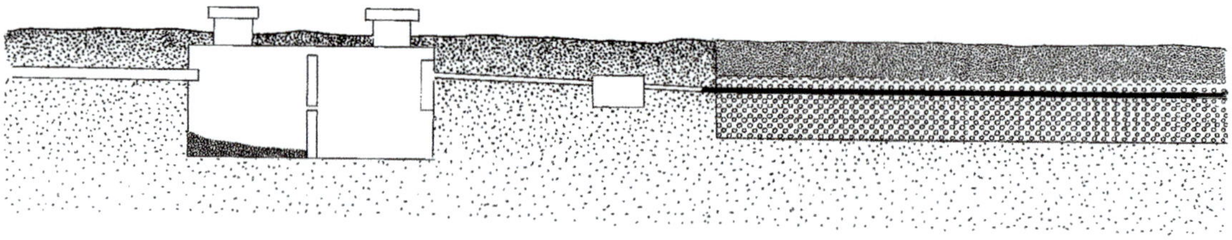

Figure 4.21 Typical configuration of a leach trench system following a septic tank (Hoover, 2013).

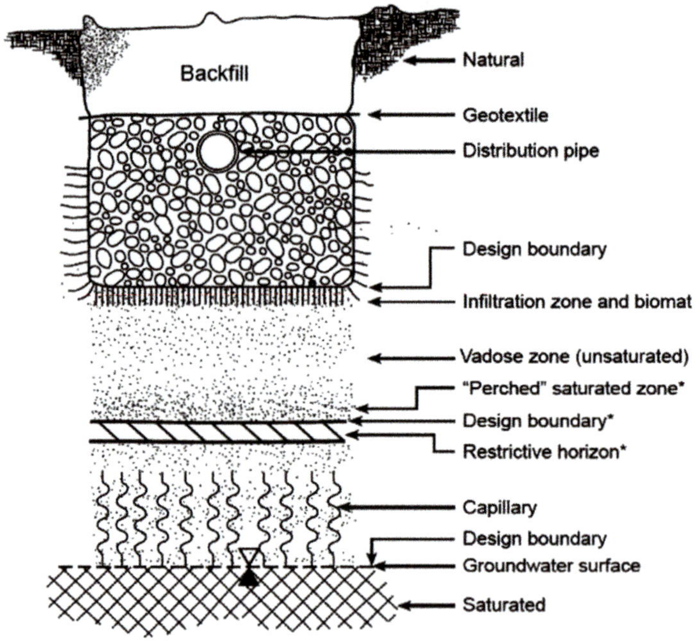

Figure 4.22 Components of leach trench and potential limiting conditions (Hoover, 2013).

Verify soils suitability by conducting the soils and site evaluation steps as described in Chapter 3 to determine that:

- a proper long-term acceptance rate (LTAR) can be calculated;
- soil depth is sufficient to maintain a separation of 0.30 m below the bottom of the trench and the shallower of either the highest level of groundwater or an impermeable zone such as bedrock or a clay horizon; and
- there is enough surface area at the site to accommodate the required size of the leach trenches or beds.

4.6.1.1.1 Design Strategy for Leach Trenches

The strategy for designing leaching systems is to first determine the required infiltrative surface area (SA), which is the sidewall (not bottom) area of the leach trenches and the bottom area of leaching beds, and then to determine the best configuration of the leach field. The variables, formulas, and a step-by-step design process based on an example are given below.

- flow (Q) of the influent to be applied to the soil, given in L/d;
- Long-term acceptance rate (LTAR), which as discussed in Chapter 3, quantifies how permeable the soil is, specifically regarding how much influent it can safely accept (given in $L/m^2/day$); and
- depth of soil (D) above the shallowest limiting condition, given in meters.

Formula 1: $SA = Q/LTAR$.

Formula 2: Trench length = SA/2 (remember that there are two sidewalls for each trench).

Example: A level site is known to have a soil depth of 2.05 m above the height of seasonally high groundwater. The LTAR has been determined by soils evaluation to be 20 $L/m^2/d$. The source flow is 7500 L/d. Assume that the septic tank will be located 15 m from the building, that the leach field will be 10 m from the septic tank, and that the building sewer exits the building at a depth of 0.25 m below the surface. Determine the following:

(1) the effective depth of the leach trench;
(2) the required length of the leach trench;
(3) the optimum configuration of the leach field assuming rectangular shape, maximum individual trench length of 30 m, separation between trenches of 2 m, and trench width of 1 m.

(1) *Effective depth*:
   (a) *Determine the maximum depth of the leach trench* – if the soil depth above seasonally high groundwater is 2.05 m, the maximum depth of the trench is 2.05 m minus 0.30 m (required separation from the bottom of the trench to the limiting feature), which gives 1.75 m maximum trench depth.
   (b) *Determine the depth of the leach pipe in the trench* – there are four measurements to consider:
   - drop from ground level to the building sewer exit point, which is 0.25 m below the surface;
   - drop from the building sewer exit point to the septic tank = distance (m) ∗ 2 percent sewer slope = 10 m ∗ 0.02 = 0.20 m;
   - drop through the (typical) septic tank = 5 cm or 0.05 m;
   - drop from the septic tank to the leach trench = distance (m) ∗ 2 percent sewer slope = 15 m ∗ 0.02 = 0.30 m.

   The sum of the drops from the four factors above = 0.25 m + 0.20 m + 0.05 m + 0.30 m = 0.80 m.
   (c) Determine the effective (infiltrative) depth of the leach trench:
   - maximum trench depth minus depth of leach pipe = 1.75 m – 0.80 m = 0.95 m.

(2) *Required length of trench*:
   (a) First determine the SA. $SA = Q/LTAR = 7{,}500\ L/20\ L/m^2/d = 375\ m^2$.
   (b) Determine the length (L) of trench. $L = SA/\text{effective depth}/2$.
   Total Length of Trench = $375\ m^2/0.95\ m/2$ sidewalls = 197 m.

(3) *Optimum configuration*:
   (a) First, determine how many trenches are required:
   - total length required divided by max trench length = 197 m/30 m = 6.6. Since the number of trenches must be a whole number, round up. Total number of trenches required = 7;
   - calculate individual trench length = 197 m/7 = 28 m. In summary, the system will require 7 trenches at 28 m each.

   (b) Now, determine the total area of the leaching system:
   - 7 trenches, each are 1 m wide with 2 m separation, gives a sum of 19 m total width of leach field;
   - total area = trench length ∗ width of leach field = 28 m ∗ 19 m = 532 m^2.

4.6.1.2 Alternative Non-mechanical Leaching Systems Similar to Trenches

Leaching chambers: Leaching chambers (Figure 4.23) are suited for sites where soils are good for on-site leaching but gravel is not readily available. The lightweight easy-to-manage units 'snap' together for rapid installation. They are placed in the upper reaches of the soil profile to maximize oxygenation of the effluent from the air in the soil pore spaces, and from evapotranspiration from plant roots. Because of their greater efficiency compared to gravel filled leach trenches, chamber-based leaching systems may require smaller areas. Check with manufactures for specific design guidance.

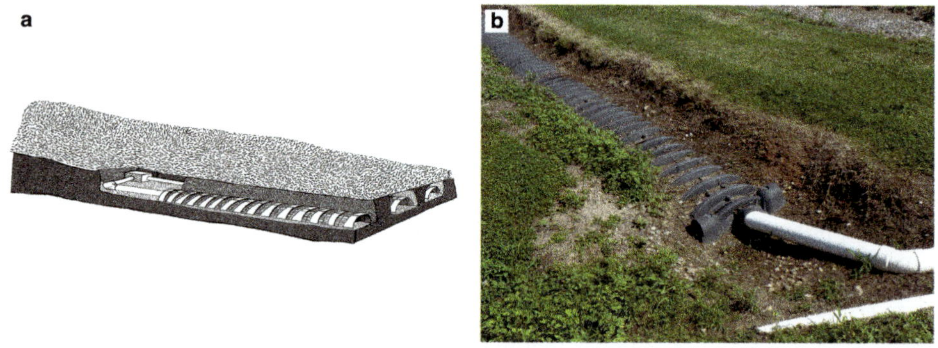

Figure 4.23 (a) Leaching chamber (Hoover, 2013); (b) actual installation of shallow leaching chambers prior to final cover (Authors).

For more technical and in-depth information on leach chambers, refer to US EPA (2000c) Decentralized Systems Technology Factsheet: Septic Tank Leaching Chamber: http://water.epa.gov/scitech/wastetech/upload/2002_06_28_mtb_septic_tank_leaching_chamber.pdf.

Leaching Beds: Unlike leach trenches, which are sized based on side-wall area, leach beds (Figure 4.24) are sized based on bottom area. Leach beds may be more advantageous for areas where soils are relatively shallow. Take care during installation to ensure that water does not pond on top by gently sloping the surface of the bed to shed storm water. It is also critical to take care during installation to ensure against compaction of the bottom of the bed. Minimize walking (and prohibit heavy equipment from sitting) on recently excavated beds, and never excavate during wet conditions.

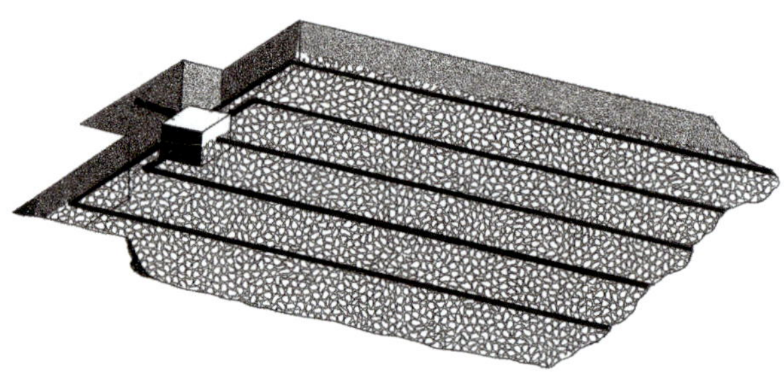

Figure 4.24 Leaching bed (Hoover, 2013).

4.6.1.2.1 Leaching Systems' Need for Even Influent Distribution

In general, one of the keys to the long-term function of leaching and other soil dispersal systems is the method of even distribution used to ensure that each trench or section of the bed receives equal amounts of influent. This is best achieved by the use of a distribution box (D Box), described below and shown in Figure 4.25. While there are many styles of distribution boxes, those constructed from plastic or concrete are most typical. For larger systems, distribution boxes can be arranged in series to allow for even dosing of larger numbers of trenches.

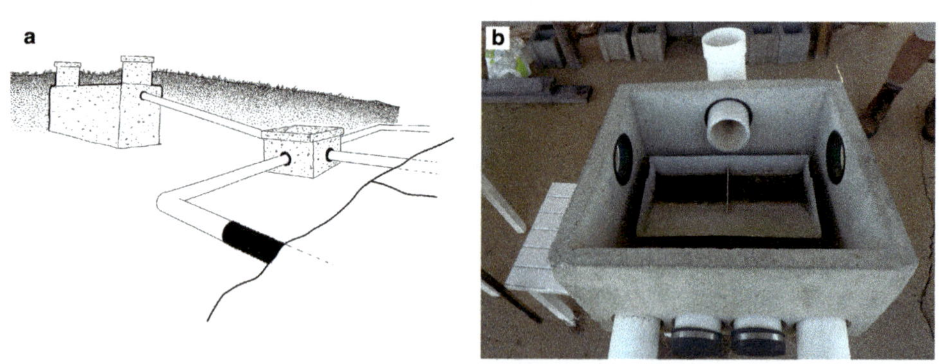

Figure 4.25 (a) D Box (Hoover, 2013); (b) D Box with tipping pan used to achieve even distribution (Authors).

Installation guidance for D Boxes

- build a wooden form around the D Box and anchor it into the earth;
- add a few inches of concrete in the form and set the D Box on top at the proper elevation;
- fill the D Box with water and level it so that all outlets are at the same elevation. Allow the concrete to set with the box in the level position;
- add more concrete inside the form to secure the D Box in the level position; and
- add speed levelers to improve even distribution even further.

4.6.1.3 Mound Systems

Mound systems (Figure 4.26) may be thought of as raised leach beds. They are often used for sites with high groundwater or where soils have inadequate permeability. Septic tank effluent is pumped to the mound which is constructed using uniform large grain sand. The sand provides both biological filtration (through the microbes that are attached to the sand grains) and physical filtration. The sand also helps to encourage evapotranspiration by the mechanism of capillary action, which wicks the effluent upwards through the sand towards the mound surface as one method of dispersal, the other being downward drainage.

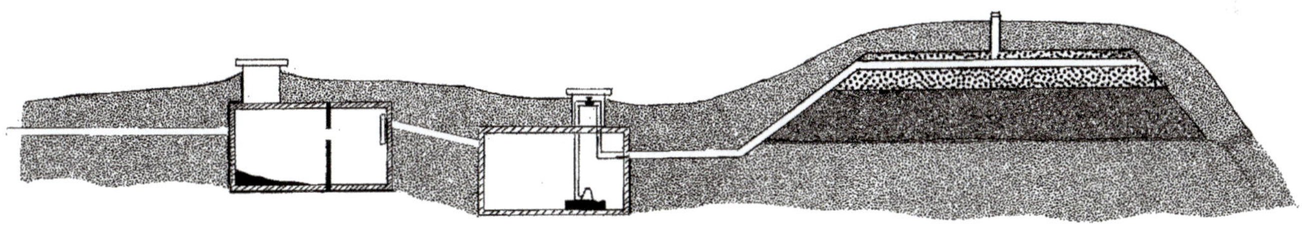

Figure 4.26 Mound system treating septic tank effluent through an elevated sand bed (Hoover 2013).

Mound systems are suited for small flows in areas where a good source of clean and uniform sand (as well as electricity for pumping) is readily available. While the area for the mound must be relatively level, if the septic tank is located upslope of the mound, the system may be designed using a non-electric dosing siphon instead of a pump system.

For more technical and in-depth information on mound systems, refer to US EPA (1999a) Technology Fact Sheet for Mound Systems: http://www.epa.gov/npdes/pubs/mound.pdf.

4.6.1.4 Drip Irrigation Systems

Drip irrigation is an effluent reuse and dispersal technology that has been proven to be a very effective and low-cost method of directly reusing wastewater effluent for watering plants. It has been used successfully for growing landscape plants, turf grasses, and fruit and nut trees. It functions by distributing small drops of effluent directly to the root zone of plants, minimizing the chances for evaporation, and thereby conserving this vital resource. As wastewater effluent contains nitrogen, phosphorus and micronutrients, it is also an ideal supplement to chemical fertilizers.

Drip irrigation systems for effluent are designed differently from drip systems using potable water. As effluent is biologically active, a simple mechanism for periodically cleaning out the drip tube is needed. For effluent systems, this is accomplished by laying out the drip tube in closed loop configurations. This is seen in Figure 4.27a, which is a demonstration of the way the system looks underground. Under operating conditions, a valve at the end of the loop is closed, which forces the effluent out through the small emitters embedded in the drip tube. When it is time to flush, the valve is opened and the effluent is forced through the tube at high velocity, effectively cleaning the tubes.

Drip irrigation systems for wastewater effluent are typically mechanized systems that utilize the same type of effluent pumping configurations as described for pressure sewers. They often use a series of solenoid valves (Figure 4.27b) that controls effluent delivery to different zones, and even does the flushing procedures automatically. The drip irrigation tube is typically installed at a depth of 15–30 cm, which minimizes excavation costs. Gravel, which is required in large quantities for leach trenches, is not used in drip fields, so material expenses can be much lower. The reduced cost of installation and the value associated with effluent reuse may make up for the cost of the pumping equipment.

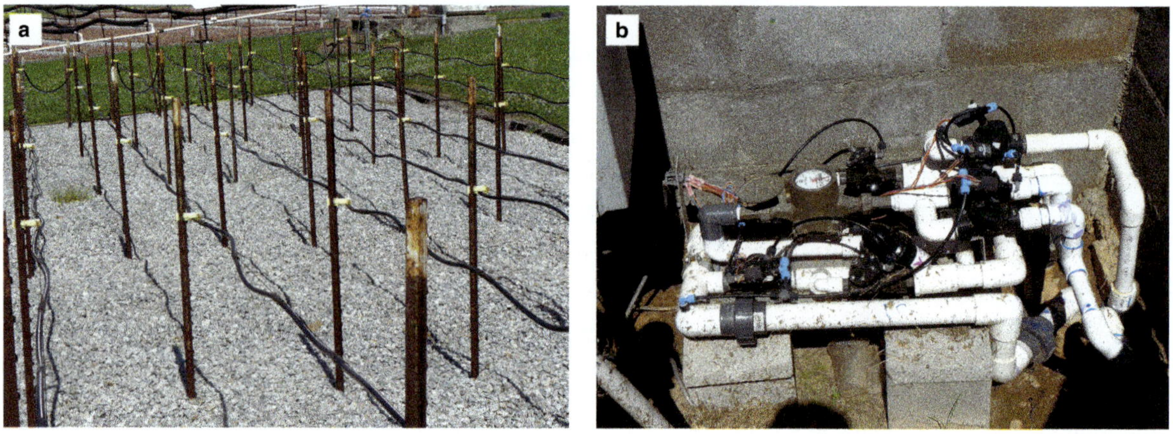

Figure 4.27 (a) Above ground drip field for demonstration at NC State University, Raleigh, U.S., with loops at the end of trenches (Authors); (b) Drip field headworks, showing solenoid valves that control distribution (Authors).

4.6.2 Constructed Wetlands

Constructed wetlands (CWs) are lined basins filled with gravel media that support plants and microbes. Those combine to retain and partially degrade organic wastes anaerobically and aerobically through microbial activity. The three types of CWs include free-water surface flow CWs (Figure 4.28), horizontal subsurface flow CWs (Figure 4.29), and vertical flow CWs (Figure 4.30), all of which achieve some removal of solids and organic matter, pathogen reduction, and potentially some nutrient uptake through the plant roots.

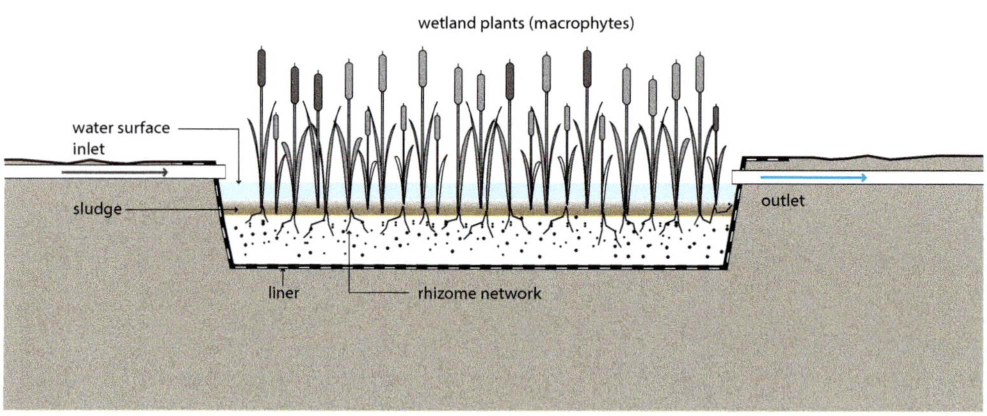

Figure 4.28 Free water surface flow constructed wetland schematic (Compendium, 2014).

The effective treatment efficiency and pathogen reductions of CWs are counterbalanced by their moderate capital and O&M costs and moderate space requirements, expert construction and O&M requirements, which can all be significant limiting factors in the application of this technology. However, CWs provide additional functions and values that may appeal at the local level, thanks to the greenery they produce. Initiated by placing around two clods of plants or four growing rhizomes per square meter of surface area (Sasse, 1998), properly cultivating and managing the vegetation can develop the unit into a beautiful green space, as well as for habitat for birds and small animals.

The main difference between the three CW configurations relates to each one's method of influent flow within the man-made basin. Free-water surface flow CWs, generally more applicable to larger flows, expose the wastewater to the air and sunlight. Horizontal subsurface flow CWs force the influent down a slightly sloped path through a planted gravel-filled basin, and are more efficient at removing organic matter, require smaller space and have fewer mosquito and odor nuisances.

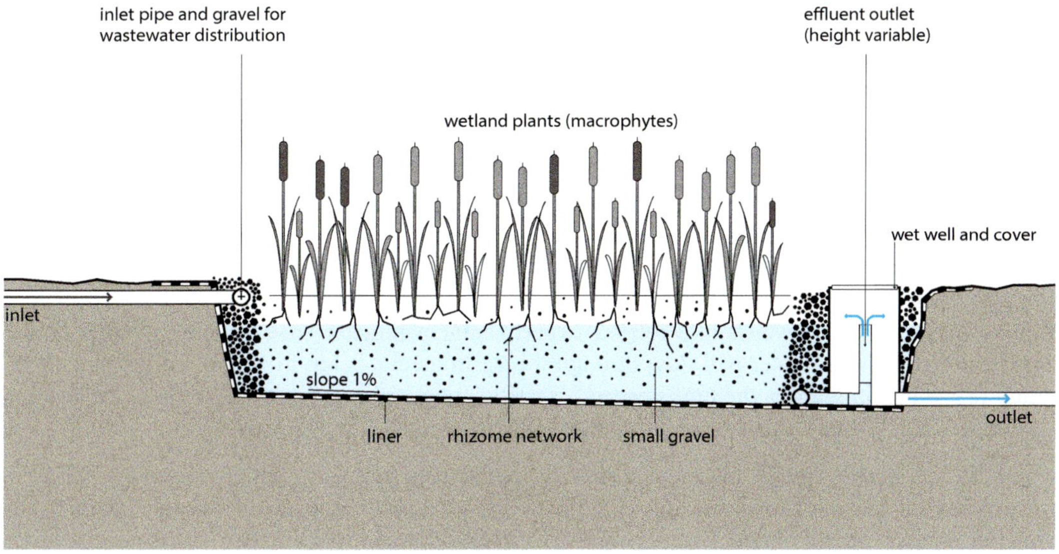

Figure 4.29 Horizontal subsurface flow constructed wetland schematic (Compendium, 2014).

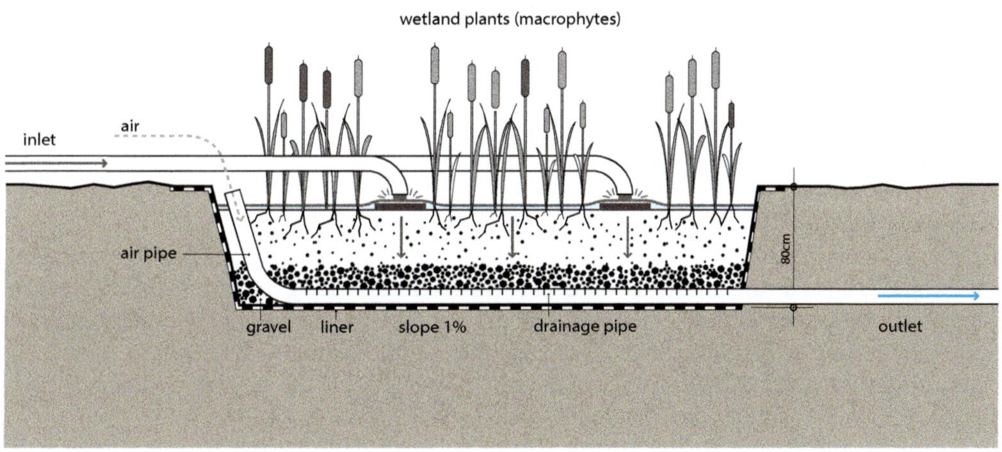

Figure 4.30 Vertical subsurface flow constructed wetland schematic (Compendium, 2014).

Vertical flow CWs differ from their horizontal subsurface flow counterparts not only in their direction of flow, but also in the manner in which they are dosed. Vertical flow CWs are often designed for intermittent dosing, achieved through use of a small effluent pump located in the last compartment of the prior septic or dosing tank. Dosing occurs through influent loading around 5 to 10 times daily, which helps entrain oxygen within the media to enhance aerobic degradation of pollutants. In this way, vertical flow CWs can accept higher organic loads than horizontal subsurface flow CWs. However, unlike the other CW types, the complex dosing of the vertical flow wetlands requires electricity and expert engineering and construction supervision.

While each of the three CW types can be effective at treating wastewater, this manual focuses mainly on the horizontal subsurface flow type. This is due to the fact that it is the most practical and nuisance-free configuration used for moderate removal of BOD, TSS, and pathogens from primary effluent. Free-water surface and vertical subsurface models have specific uses that may be explored further in the references.

For basic overview of the three types of CWs, refer to the Compendium, pp 114–119. For more technical and in-depth information, refer to US EPA (2000d) Wastewater Technology Fact Sheet on Free Water Surface Wetlands: http://water.epa.gov/scitech/wastetech/upload/2002_06_28_mtb_free_water_surface_wetlands.pdf, and US EPA (2000e) Wastewater Technology Fact Sheet on Wetlands, (Horizontal) Subsurface Flow: http://water.epa.gov/scitech/wastetech/upload/2002_06_28_mtb_wetlands-subsurface_flow.pdf.

4.6.2.1 Design Strategy for Horizontal Subsurface Flow Constructed Wetlands

Note: Some of these formulas and calculations may appear to be difficult or confusing. The toolkit described below helps users apply these equations for their horizontal subsurface flow constructed wetland (HSCW) projects. The procedure and associated math, concluded with an instructive example with full calculations, are in the sections below.

The sizing and configuration of HSCWs is mostly determined by the following variables:

- flow (Q), which is the daily volumetric loading rate of the influent, given in m^3/d;
- organic loading for the influent (BOD_i) and also for the effluent standard (BOD_e), given in mg/L;
- porosity (P), which is a proportion that indicates the amount of space between the gravel particles that is filled with the flowing wastewater;
- hydraulic conductivity (HC), which is a related experimental parameter based on the average gravel diameter. It represents how fast the wastewater will flow through the basin media, given in simplified units of m/d;
- temperature (T) which influences the time needed for sufficient treatment, given in degrees Celsius; and
- depth (D) of the HSCW basin that can be dug based on soil conditions, given in meters.

Several mathematical relationships guide the calculation of proper surface area (SA), which makes the HSCW big enough for sufficient treatment, and of cross sectional area (CSA), which must be large enough to allow all influent to flow under the surface. Full references to these simplified formulas (UN-HABITAT, 2008) are provided at the end of this section. These resources have been combined into the following step-by-step procedure that helps simplify the calculations:

(1) Choose the preferred depth, to be excavated uniformly throughout the bed, at a site with a relatively flat slope. D should be between 0.4 and 0.8 m; towards the upper end of that range is best. With a safety buffer between the bottom of the lined basin and 0.3 m added in, that range becomes 0.7–1.1 m, and represents the total necessary depth to limiting condition. Assuming there is no soil limiting condition at a depth shallower than that range, D is determined by the type of plant that will be utilized and its root depth. Bulrushes (*Scirpus*) and reeds (*Phragmites*) are two of the most effective treatment plants, with root depths of 0.8 and 0.6 m, respectively.

(2) Select the preferred (ideally, also the most-available) gravel media. The gravel should be as uniform as possible, free of fines, and with a diameter between 5 and 32 mm.

(3) Determine the porosity of the gravel. It is found by filling a bucket or smaller graduated container with a representative sample of the gravel up to a volume marker line, such as a 5 L line, making sure that the surface is flat at the line. Next, fill the bucket with water until it reaches the rock volume line. Finally, pour all of that water into another bucket or any other receptacle, as long as it has precise volume markings for estimating volume within around 10 mL, such as one circumference line for every 50 mL. Divide the amount of water by the original volume, to get the fraction that represents P. HSCWs work best when that value is between 0.35 and 0.40.

(4) Calculate SA based on the plug-flow BOD removal equation Formula 1 shown below. This requires that the user inputs values for BOD for influent and effluent (as determined by local standards), as well as for flow, temperature, porosity and depth. Note that in cold climates, T must be decreased slightly, as the HSCW bed winter temperature will be much lower than its annual average value due to significant heat loss.

$$\text{Formula 1:}\quad \text{SA} = \frac{Q * (\ln|\text{BOD influent}| - \ln|\text{BOD effluent}|)}{P * D * 1.104 * 1.06^\wedge(T - 20)}$$

(5) Use Formula 2 below to calculate the necessary width (a determinant of CSA) to ensure a sufficient subsurface flow rate (Q_S), which is set to equal Q. If Q_s is less than Q, surface overflow would occur, which is the main condition that must be avoided at all costs for HSCWs. This equation requires values for HC, D and hydraulic gradient (HG), which is the drop in the water surface height from the inlet to the outlet. It is taken to be 0.01.

$$\text{Formula 2:}\quad W = \frac{Q}{\text{HC} * 0.01 * D}$$

(6) Complete the math by calculating individual cell dimensions. Once W is determined, the total SA may be partitioned into smaller cells to ensure proper length to width ratios (not more than 3:1 and not less than 0.4:1).This will depend upon the nature of the site and the configuration of the available land area.

(7) Consult a site engineer to determine the ideal layout of the cells, given that space restrictions may not be favorable to one long row (width-wise) of parallel cells.

Example: A small rural residential subdivision of 50 homes, with an average of four people per home and typical wastewater production rate of 200 L/person/d, produces a flow rate Q of 40 m^3/d. The average BOD influent level from several samples as measured at a local laboratory is 250 mg/L, and the local effluent standards require 20 mg/L or less for discharge. The annual mean temperature of the area is 25 degrees Celsius. The project developers conduct the five step process as follows:

(1) They search for a relatively flat site with sufficient depth to hard rock or groundwater, preferably at least 0.6 m deep to allow for the planting of bulrushes or reeds. A site with a depth of 0.6 m, appropriate for their abundant local reed species, is chosen.
(2) They search for a large supply of cheap 5 to 32 mm-wide gravel, and they decide on using locally abundant 16 mm fine gravel.
(3) They fill a 5 L bucket with the 16 mm gravel and top it off with water. Pouring out the water reveals a pore water volume of 1.9 L. This is divided by 5 L to determine the gravel's porosity as 0.38.
(4) Using the HSCW interactive tool, they plug their data into the tool (described below) input menu, which first uses Formula 1. This determines that their HSCW surface area should be 300 m^2.
(5) The tool automatically finds the necessary width to be 50 m.
(6) The tool then divides that width of 50 by 15 m to get 3.3 cells and rounds that up to 4. The 50 m width is then divided by 4 to determine a width per cell of 12.5 m. Length is found by dividing the surface area of 300 m^2 by the overall width of 50 m to get a length of 6 m.
(7) The site engineer presents the two simplest 6 by 12.5 m cell configuration options to the developer – either a one row setup of four parallel cells, requiring a 6 by 50 m space, or two adjacent rows of two cells each, requiring a 12 by 25 m space.

Access the HSCW design toolkit and site with details on the variables and other key aspects here: http://www.iwawaterwiki.org/xwiki/bin/view/Articles/WastewaterSolutionsDesignToolsforEngineers.

Constructed wetlands guidance related to source characteristics and site limitations

- *influent flow rate:* although CWs can be scaled up to manage high flows, that is usually not feasible in most settings, as the increasing land requirements can create siting and/or cost barriers. For the purposes of conservative guidance, constructed wetlands may be considered as viable options for sources with flow rates below 100 m^3/d, although much greater flows have been managed by CW systems for small municipalities and individual utilities that can provide the high level of O&M and desludging required (Gauss, 2008);
- *influent strength:* CWs are subject to clogging that can require full replacement of gravel, so be sure to verify the concentration of FOG, solids and organic matter. While in some cases high BOD influent may be acceptable for short periods of time, an upper limit of 170 mg/L, which is the average level for primary-treated sewage influent entering HSCWs recorded in a recent review of hundreds of the units (Vymazal & Kröpfelová, 2009), is appropriate to ensure sustained functioning. If the influent strength exceeds that level, utilize additional appropriate pretreatment and primary treatment devices to reduce influent strength, as is demonstrated in the toolkit;
- *slope:* CWs require moderate land spaces that are relatively level, with a 1 percent slope being ideal. Avoid using the technology for sites with slopes greater than this that cannot be cost-effectively excavated or backfilled to ensure that slope;
- *local land use restrictions:* nearby parcels, such as airports, may not permit the installation of such natural basins, which can cause problems due to odors, mosquitoes, or an excess of birds or other wildlife;
- *other considerations:* site-specific issues related to available gravel sources and climates that are conducive to mosquito and weed growth can be causes for concern. Make sure that obtaining and installing the gravel is not only cost-effective, but also done correctly, as it can be contaminated with fines, dirt, and debris. Use prewashed gravel if possible to avoid this problem.

4.6.3 Waste Stabilization Ponds

Waste stabilization ponds (WSPs) are lined ponds or basins that treat wastewater through the physical action of settling and the biological action of microbial digestion. Also known as sewage lagoons, they are popular because they are relatively inexpensive and easy to design, install and maintain and can produce good quality effluent. They do, however, require large land areas.

In many applications, WSP systems use multiple units together to achieve primary, secondary and tertiary treatment. This stepwise series is shown in Figure 4.31. However, in some cases aerated ponds can be utilized within or outside the series. While these systems are frequently used for community wastewater management, in general they are valid DWMS options (in series and apart) that are worthy of consideration for many applications including serving as the main treatment component within septage management programs.

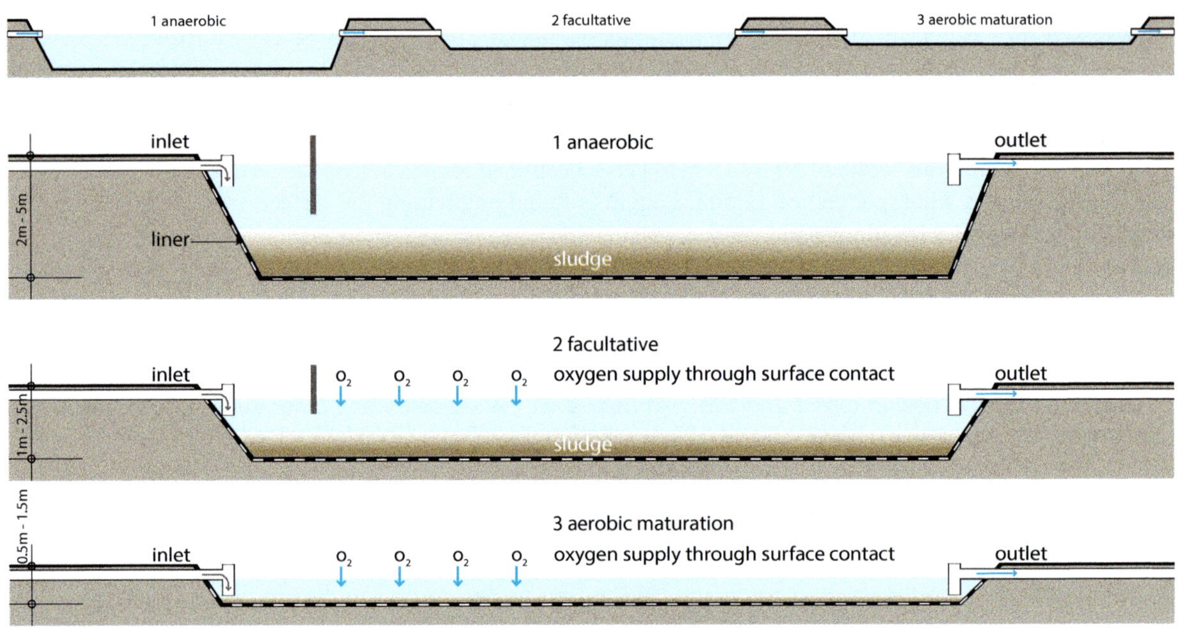

Figure 4.31 Typical waste stabilization pond series (Compendium, 2014).

For basic overview of this basic WSP scheme, refer to the Compendium, pp. 110–111.

4.6.3.1 Anaerobic Ponds

In general, anaerobic ponds (AnPs) can be considered as a primary treatment step, as they anaerobically degrade and settle organic matter and solids. However, they are presented here in the secondary treatment section due to their frequent stepwise linkage (as seen in Figure 4.32a and b) with the other two major types of WSPs, facultative and aerobic ponds, which are the secondary and tertiary treatment stages, respectively. This is not to say that AnPs, as a highly effective primary treatment mechanism, cannot be used by themselves in sequence with non-pond (secondary and tertiary) treatment steps. In fact, this scenario is often warranted given space requirements, as AnPs require less land for a given flow than do the other pond types, which makes them more practical when finding sufficient land space becomes an issue.

In general, AnPs are deliberately not aerated, heated, or mixed and have deep basins from 2 to 4 m or even 5 m deep (Compendium, 2014); deeper is better. These characteristics, particularly the depth, do not permit oxygen to reach the bottom zone, allowing anaerobic conditions to dominate. In this respect, AnPs are different from shallower facultative or aerobic ponds, as those conditions are similar to those occurring within an unheated anaerobic digester. However, conventional digesters are typically used for sludge stabilization, whereas pond systems typically are used for both primary and secondary wastewater treatment. That being said, AnPs and the following facultative ponds may be designed specifically for fecal sludge treatment, which is illustrated in the case study from Dumaguete City, The Philippines, presented below.

Figure 4.32 (a) Properly-desludged, well-maintained anaerobic pond within a system of all three pond types, some in series and others in parallel, serving a large community in Arani, Bolivia (E. Symonds, personal communication, December 14, 2013); (b) Poorly-maintained anaerobic pond, lacking a proper grit removal chamber and desludging services, within a similar pond series serving a small town in the Cochabamba Valley of Bolivia (Verbyla, 2013).

For more technical and in-depth information, refer to US EPA (2002b) Wastewater Technology Fact Sheet on Anaerobic Lagoons: http://www.epa.gov/owm/septic/pubs/alagoons.pdf.

4.6.3.2 Facultative Ponds

As the next step in the traditional WSP series, facultative waste stabilization ponds (FPs, shown in Figure 4.33a) are used mainly for secondary treatment for municipal and industrial wastewater. These lined earthen ponds are usually excavated to depths between 1 and 2.5 m (Compendium 2014) and are neither mechanically mixed nor aerated. The layer of water near the surface contains dissolved oxygen due to atmospheric re-aeration and algal respiration, which supports aerobic and facultative organisms. Sludge deposits accumulate at the bottom of the pond, where anaerobic organisms flourish. The middle layer, termed the facultative zone, ranges from aerobic near the top to anaerobic at the bottom and supports microbial degradation of organics and pathogens. As with AnPs, they can also be used in sequence with non-pond (primary and tertiary) treatment technologies.

Figure 4.33 (a) Facultative pond serving 780 people in Alto Beni, Bolivia (Verbyla, 2013); (b) Two aerobic ponds in series treating the effluent of a UASB reactor serving 1,300 people in Alto Beni, Bolivia (Verbyla, 2013).

For more technical and in-depth information, refer to US EPA (2002c) Wastewater Technology Fact Sheet on Facultative Lagoons: http://www.epa.gov/owm/septic/pubs/faclagon.pdf.

4.6.3.3 Aerobic Ponds

Aerobic ponds (APs, shown in Figure 4.33b) complete the WSP series by providing final polishing through reducing pathogens and nutrient levels. Also known as maturation or polishing ponds, they are shallow basins typically between 0.5 and 1.5 m deep (Compendium, 2014). Dissolved oxygen concentrations are usually high due to wind and wave action that allow for natural re-aeration, which combined with the effects of the UV radiation (which itself directly impairs pathogens' internal functioning when the water column is so shallow) stimulates algal growth. The algae change the pH of the water outside of the tolerable pH range of most pathogens and take up much of the phosphorus and nitrogen if the algae are harvested or consumed by fish in aquaculture. Otherwise, those nutrients would likely be discharged and could pose a threat to surrounding aquatic habitats. As with the above two pond types, they can also be used in sequence with non-pond (primary and secondary) treatment steps.

For more detailed discussion, refer to Sasse (1998), pp. 99–103.

4.6.3.3.1 Design Strategy

Note: Some of these formulas and calculations may appear to be difficult or confusing. The authors toolkit described below helps users apply these equations for their WSP projects. The procedure and associated math, concluded with an instructive example with full calculations, are in the sections below.

The various configurations of WSP series are generally based on the initial influent strength. If the BOD of the influent wastewater is moderately high – for these purposes, at least 400 mg/L – the first pond in the series (parallel pond-type systems are not yet considered within the toolkit) will be anaerobic; otherwise, it will be an FP. If BOD is very high, one or more AnPs will be needed to reduce BOD sufficiently before an FP can receive the wastewater. In all cases, those are followed by one or more MPs to achieve sufficient pathogen removal. In general, the designation of pond series and calculations for each pond's sizing are dependent on several variables, all of which are described below:

- flow (Q), which is the daily volumetric loading rate of the influent, given in m^3/d;
- organic loading, given in mg/L, for the influent (BOD_i) and also for effluent standard (BOD_e), which is taken to be at most 50 mg/L;
- net evaporation (E), which is water loss from evaporation minus rainfall, given in mm/year;
- temperature (T), given in degrees Celsius, which influences BOD volumetric and surface loading rates;
- depth (D) of the overall site that can be dug based on soil conditions, given in meters.

Given that the literature has not settled on a convention for setting pond series based on BOD_i or COD_i, the authors have proposed a system, described below, whose principles are best explained by an example. If BOD_i is 600 mg/L, following the formulas explained shortly will give BOD_e of around 37.5 using an AnP-FP-AP pond series, while using the same series for a setting with BOD_i of 300 will give BOD_e of 18.75.

Since this is unnecessarily low based on the 50 mg/L BOD_e assumed standard, the AnP pond can be removed, which would allow the FP to serve as the primary treatment mechanism and gives BOD_e of 37.5 as preferred. This phenomenon leads to the preferential selection of the following pond series based on BOD_i, all based on the flexible assumption that BOD_e cannot exceed 50 mg/L:

- for BOD_i greater than 4,444 mg/L, a five pond series of AnP-AnP-FP-AP-AP is used;
- for BOD_i greater than 1,333 mg/L, a four pond series of AnP-AnP-FP-AP is used;
- for BOD_i greater than 400 mg/L, a three pond series of AnP-FP-AP is used;
- for BOD_i less than or equal to 400 mg/L, a two pond series of FP-AP is used.

Full references to the below formulas (Kayombo *et al.*, 2005) and this text are provided at the end of this section. These resources have been synthesized into the following step-by-step procedure that simplifies calculations:

(1) After determining Q, BOD_i, BOD_e, T, E and D, BOD_i is used to set the pond series.
(2) Sizing calculations for the one or two AnPs (if needed) are undertaken. The AP BOD volumetric loading rate (VLR_{BOD}) is determined based on T using a simple step-wise function that assumes a minimum VLR_{BOD} of $100\ g/m^3$ for $T < 10°C$ and $350\ g/m^3$ for $T > 25°C$. Along with flow and BOD_i, it is used in Formula 1 shown below. This calculates AnP volume, which divided by depth gives AnP surface area. Knowing that for WSPs, the

length of each pond can be taken as three times that of its width, it is simple to find the AnP (and other ponds') length and width, which equals the square root of the 'surface area divided by 3'. BOD_e for the AnP, which becomes BOD_i for the following AnP or FP, is found by multiplying BOD_i by a value ranging from 0.60 (40 percent BOD reduction) for $T < 10°C$, and 0.30 (70 percent BOD reduction) for $T > 25$. The values within this range of 0.30 to 0.60 are obtained from a simple step-wise function solely dependent on T.

$$\text{Formula 1: AnP volume} = \frac{\text{BOD}_\text{i} * Q}{\text{VLR}}$$

(3) Sizing calculations for the single FP are undertaken. The FP BOD surface area loading rate ($SALR_{BOD}$) is determined based on T using a function labeled Formula 2, shown below. It, along with BOD_i (BOD_e in Step 2) and Q, is used to calculate FP surface area (Formula 3, below), which can be multiplied by D (taken as 2.1 m) to get FP volume. BOD_e is found by multiplying BOD_i by the assumed removal factor of 75 percent.

$$\text{Formula 2: FP SALR}_\text{BOD} = 350 * (1.107 - (0.002 * T))^{(T-20)}$$

$$\text{Formula 3: FP surface area} = \frac{10 * \text{BOD}_\text{i} * Q}{\text{SALR}}$$

(4) Sizing calculations for the one or two MPs are undertaken. MP surface area is based on Q, D (taken as 1.3 m), E and hydraulic retention time (HRT). The latter term, utilized in the equations for other DWMS technology sizing, is assumed to be five days for the purposes of this tool. Formula 4, shown below, demonstrates how MP is related to those variables.

$$\text{Formula 4: MP surface area} = \frac{2 * Q * \text{HRT}}{2 * D + 0.001 * \frac{E}{365}}$$

Example: a hypothetical scenario for a potential large meat-focused public market project is given. The market managers decided to follow step-by-step planning process to determine the potential suitability of WSPs. The process went as follows:

(1) They recruited a local engineer for data collection, who found that average Q was 50 m^3/d, and that average discharged organic loading (BOD_i) was around 1,600 mg/L. This BOD_i value led them within the toolkit to the four pond AP-AP-FP-MP series. The site's annual average temperature was 20°C, maximum usable soil depth was 4.8 m (which given the safety buffer soil depth of 0.3 m, meant that D was 4.5 m), and E was estimated to be 1,300 mm/yr.
(2) The toolkit used this data to estimate the AnP VLR_{BOD} for both AnPs as 300 g BOD/m^3 per day, which was used in Formula 1 with the other data to calculate volume of 266.7 m^3 and 106.7 m^3 for the first and second AnPs, respectively. Given D of 4.5 m, the tool calculated necessary surface areas of 87 and 42 m^2 for the first and second ponds, respectively. The back-to-back BOD_i reductions of 60 percent that characterized both ponds led to AnP BOD_e of 256 mg/L.
(3) This 256 mg/L value was then taken as BOD_i for the FP. That number, along with Q and FP $SALR_{BOD}$ (calculated via Formula 2 and found to be 350), was used automatically within the toolkit's Formula 3 procedure to determine that necessary FP surface area was 365.7 m^2. Multiplied by the assumed depth of 2.1 m, this gave FP volume of 707.3 m^3. This is slightly below the product of 2.1 and 365.7 because the basin has less volume than a rectangular prism with those dimensions due to inward-sloping sidewalls. Given the assumed 75 percent FP BOD reduction, this led to effluent with BOD_e of 64 mg/L.
(4) The toolkit finished the pond series sizing procedure by calculating necessary MP surface area using BOD_i of 64 mg/L (from Step 3), Q, E and assumed D and DT of 1.3 m and 5 days, respectively. That surface area value came to 191 m^2, with volume coming to 233 m^3 and final BOD_e of 32 mg/L. In summary, the tool estimated the four ponds' overall surface area (land) and volume (excavation) requirement to be 686 and 1,314 m^3, respectively, which is roughly the surface area of a square with side length of 26 m.

Access the WSP design toolkit and site with details on the variables and other key aspects here: http://www.iwawaterwiki.org/xwiki/bin/view/Articles/WastewaterSolutionsDesignToolsforEngineers#HWasteStabilizationPonds.

Waste stabilization ponds guidance related to source characteristics and site limitations

- *slope:* ponds require large land parcels that are relatively level, with a 1 to 2 percent slope being ideal. Avoid using the technology for sites with slopes greater than this that cannot be cost-effectively excavated or backfilled to ensure that slope;
- *depth above soils limiting conditions:* the placement of APs, which require depths of 2 m and function best with basin depths of around 4 to 5 m, can only be done in sites with very deep soils. This may imply the need for additional site scoping and excavation of test holes to verify that the basin depth and buffer of 0.3 m can be safely achieved;
- *local land use restrictions:* nearby parcels may not permit the installation of such natural basins, which can cause problems due to odors, mosquitoes, or excessive birds or other wildlife.

4.6.3.4 Aerated Ponds

In aerated ponds, oxygen is supplied mainly through mechanical or diffused aeration rather than by wind or wave action, or algal photosynthesis (Figure 4.34). Aerated ponds are typically classified by the amount of mixing provided. A partial-mix system provides only enough aeration to satisfy the oxygen requirements of the system and does not provide energy to keep all total suspended solids in suspension, which is required for full mix systems that have greater treatment effectiveness along with higher cost. In general, this need for electricity, although not very high at around 1 to 3 W per cubic meter of pond volume (Sasse, 1998), and the expertise needed for construction and O&M generally make this a less suitable technology for DWMS than the other pond types.

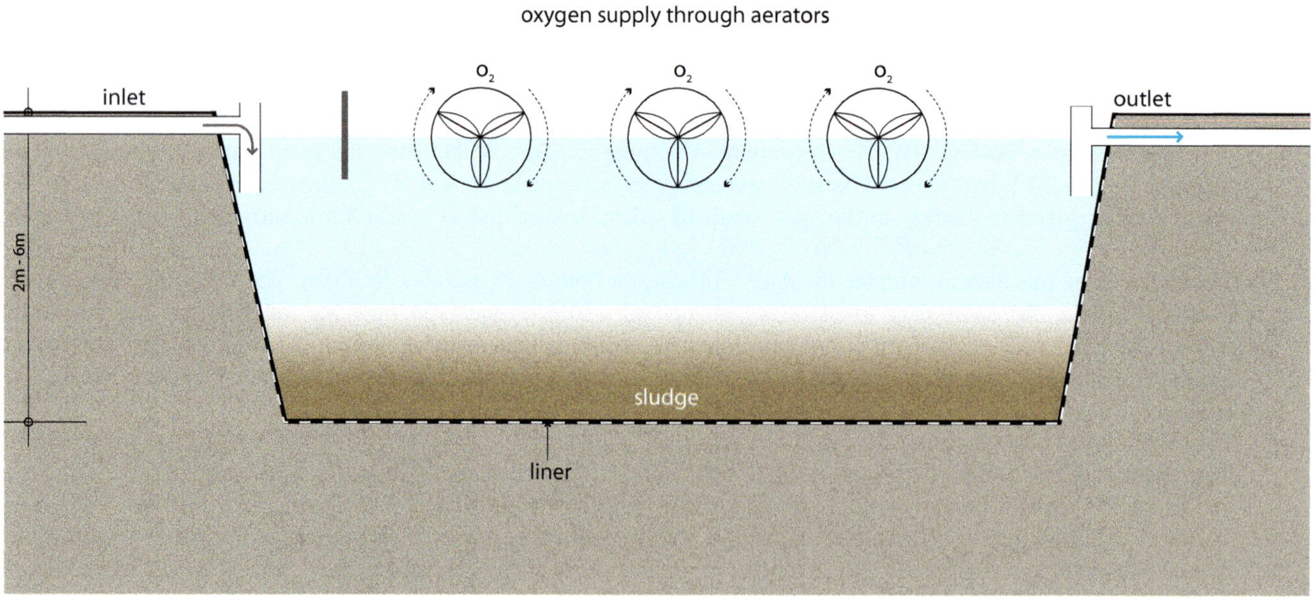

Figure 4.34 Aerated pond schematic (Compendium, 2014).

For more technical and in-depth information on Aerated Lagoons refer to US EPA (2002d) Wastewater Technology Factsheet on Aerated, Partial Mix Lagoons: http://www.epa.gov/owm/septic/pubs/apartlag.pdf.

Waste stabilization ponds used for septage management in Dumaguete, The Philippines

Dumaguete City is located on the island of Negros in the Vasayas region of The Philippines. In 2008, the city government began planning for their septage management program, which involves septic tank desludging every five years. The program is a joint effort between the city government, which designed and manages the treatment system, and the City Water District which manages the collection program.

The septage treatment system uses a series of anaerobic and facultative pond cells (Figure 4.35a), and surface and subsurface flow constructed wetlands (Figure 4.35b). Sludge drying beds are provided for dewatering the accumulated biosolids, and other than two sludge pumps, the system is totally passive. It is capable of handling a daily load of 80 cubic meters of fecal sludge per day.

Figure 4.35 (a) Josie Antonio overlooking the waste stabilization ponds (Authors); (b) The horizontal subsurface flow constructed wetland that in tandem with the surface flow wetland further treats the waste stabilization pond effluent (Authors).

'The system functions impressively well', according to Josie Antonio, City Planning Officer and principal proponent of the project. She also noted 'our last test results showed that BOD of the effluent was better than 20 mg/L [relative to BOD_i of 5,000 mg/L], indicating a very high level of treatment, while our fish living in the final indicator pond appear to be very healthy'. In order to achieve this success, it was necessary to conduct a flood study and install stream bank stabilization dykes, as the system is located on the banks of the Okoy River. Also, flow variability is high and several years were anticipated before the system was fully utilized.

More information on this case study can be found at: http://www.iwawaterwiki.org/xwiki/bin/view/Articles/Project+7+-+Country+x.

4.6.4 Media Filters

Media filters are secondary wastewater treatment devices that treat primary effluent by passing it through media such as rock, plastic or biomass. The media provide surface area where microbes become attached through a fixed film. As effluent passes over the fixed film, the microbes consume the organic matter as food. Examples of the technology include trickling filters, cocopeat filters, and sand filters. These 'fixed growth' systems are useful in land constrained areas, as they take up little space. Innovations in media filter technology are resulting in products with higher efficiency and ease of O&M, all at a reduced cost.

One common example of media filter technology is the trickling filter, which is the aerobic version of an anaerobic filter. To keep the biofilm and coarse rock supplement media partially aerated, wastewater is trickled onto the top surface while both the top and bottom of the filter are ventilated (Figure 4.36). Ensuring this requires expert facility installation and O&M, along with constant flow rate and electricity. These aspects make trickling filters feasible mostly for denser-settlement DWMS.

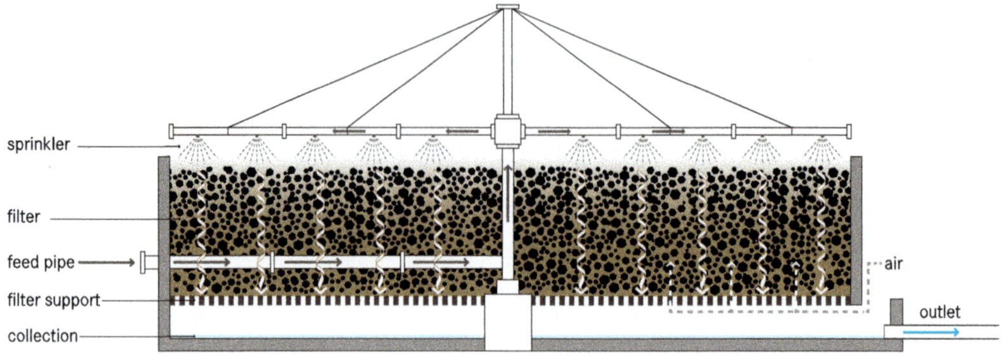

Figure 4.36 Trickling filter schematic (Compendium, 2012).

The coconut fiber (coir) and the dust that remains when the coir is removed from crushed coconut shells (cocopeat) is another biofilter medium used for secondary treatment. As shown in Figure 4.37, it is a low-cost fixed growth system that has applicability for DWMS in coconut-producing regions of the world. The main benefits of cocopeat filters as secondary treatment are their pollution reduction capabilities, lower land area needs and costs relative to CWs and WSPs, and rapid setup time. The example shown treats wastewater from 700 faculty and students for less than 1 US cent per user per day.

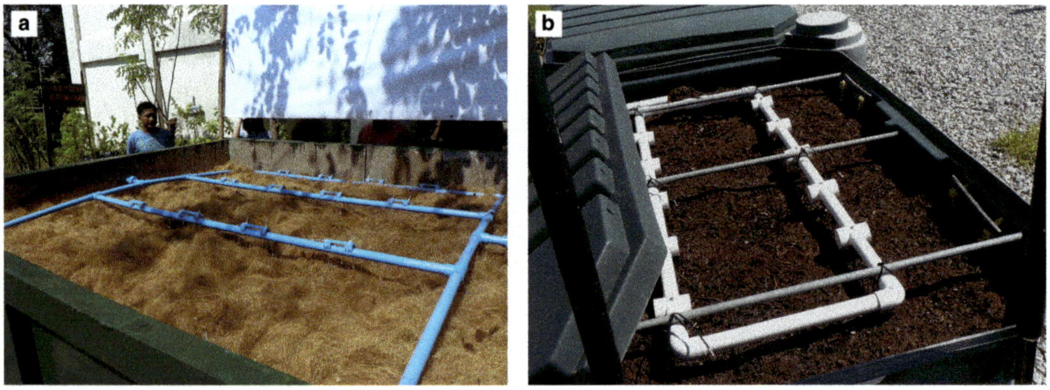

Figure 4.37 (a) Cocopeat biofilter for Putatan Elementary School, Muntinlupa City, The Philippines using pressure distribution (Authors); (b) Peat filter using gravity distribution (Authors).

Cocopeat media filters can achieve 80 percent to 90 percent reductions in pathogens, TSS and BOD for influent with BOD up to 350 mg/L. They are able to flexibly manage secondary influent from residential or industrial wastewater streams with inconsistent organic levels and flow rates, which makes them tailored for households or small institutions, although like with all media filters special focus must be put on pretreatment for FOG.

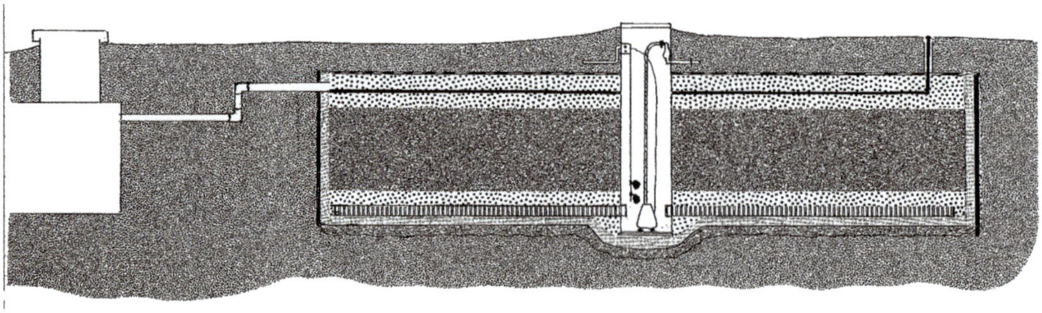

Figure 4.38 Sand filter schematic (Hoover, 2013).

Sand and gravel filters (Figure 4.38) are highly effective wastewater treatment units and have many DWMS applications. They may be intermittently dosed or designed as recirculating high rate filters (although this is still experimental). The technology has been used for years in water treatment applications, with more recent applications developed for wastewater. Media selection and proper pretreatment are keys to long-term functionality. Sand must be of uniform, large grain size, and free of fine particles. Proper grease management must be practiced, as remediating fouled media filters once they are clogged with grease requires replacement, which is expensive and time-consuming.

For a basic overview of trickling filters, refer to the Compendium, pp. 120–121. For more technical and in-depth information, refer to US EPA (2000f) Wastewater Technology Fact Sheet on Trickling Filters: http://water.epa.gov/scitech/wastetech/upload/2002_06_28_mtb_trickling_filter.pdf.
More information about cocopeat biofilter testing and treatment capability can be found at (Parker, 2011): http://www.iwawaterwiki.org/xwiki/bin/view/Articles/CocopeatBiofiltrationSystems.
For more technical and in-depth information on sand filters, refer to US EPA (1999b) Technology Fact Sheet on Intermittent Sand Filters: http://water.epa.gov/aboutow/owm/upload/2005_07_14_isf.pdf.

4.6.5 Aerobic Systems

There are many different configurations of aerobic system technology. They may use fixed growth, suspended growth or a combination of such to achieve wastewater treatment. Mostly, these technologies, except for rotating biological contractors, pump oxygen into an aeration basin to support the rapid growth of aerobic bacteria in suspension that consume and lead to high reductions in organic matter.

For DWMS application, aerobic treatment systems are most applicable to sites with limited land area and stringent local requirements for effluent quality. Four aerobic technologies are presented to provide the reader with an overview. Review the resources in the reference section for a broader discussion of aerobic technology. The four technologies presented in this manual are:

- activated sludge;
- extended aeration package plants;
- Sequencing Batch Reactors (SBRs); and
- Rotating Biological Contactors (RBCs).

The activated sludge process has been utilized for more than 130 years and represents the backbone of municipal wastewater secondary treatment around the world, as it can handle daily flows in the range of thousands to tens of thousands of cubic meters per day. As seen in Figure 4.39, it aerates and suspends influent to allow aerobic and other bacteria to degrade the organics. That bacteria-rich wastewater is then sent to a clarifying chamber, in which the bacteria clusters settle to the bottom. The more-purified wastewater is then sent to a discharge or tertiary treatment mechanism, while the sludge can be removed and partially recirculated back into the first chamber, where its bacterial clusters foster new bacterial growth and continue aerobic degradation of new influent.

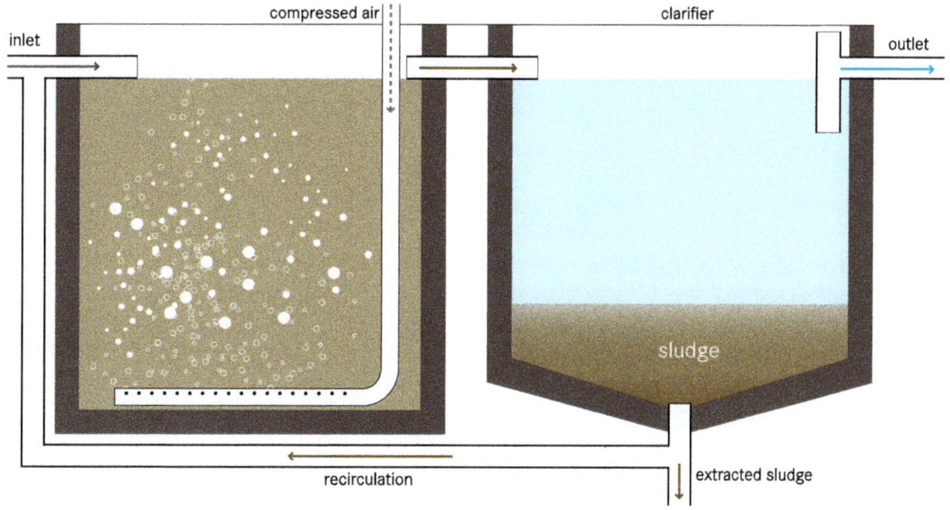

Figure 4.39 Activated sludge schematic (Compendium, 2012).

This process is often adapted for much smaller flows into a more comprehensive system, referred to as an extended aeration package plant (Figure 4.40), that may be used within DWMS for sources such as hospitals, schools or subdivisions. The plant simplifies the activated sludge process while adding various compartments for flow equalization (tanks), disinfection and sludge holding and digestion, and in some cases, for primary settling. Enclosing these within a steel or concrete tank creates an all-in-one unit that, depending on the presence of screening and primary solids settling, can in special cases serve as the only treatment mechanism for sources with high flows and low organic loading.

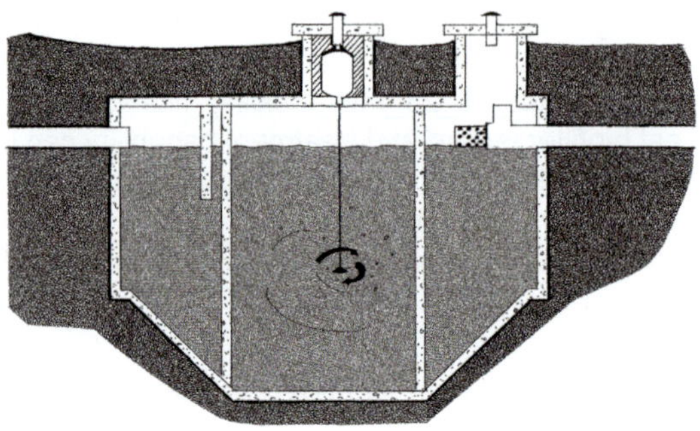

Figure 4.40 Extended aeration system schematic (Hoover, 2013).

The SBR is yet another version of the activated sludge process in which the aeration and settling processes occur in one tank, requiring less land area than the previous systems. Wastewater is processed in a variety of timed steps, which also involve flow equalization and often primary clarification. All of these facets depend on the SBR design and purpose, which generally is to treat low- or intermittent-flow sources of wastewater (US EPA, 1999c).

In contrast to the previous 'suspended growth' aerobic systems that use tanks, the RBC is an 'attached growth' system that consists of a large rotating disk with maximized surface area. That provides space for biofilm formation for microbial degradation of organics, which is enhanced by the periodic rotation of the disc into and out of water, which gives the biofilm alternating aerobic and anaerobic conditions. Most RBC equipment must be imported, likely at great expense, while like most other mechanized secondary treatment systems, they require expert installation and O&M along with a continuous source of power. All of these aspects make RBCs applicable mostly for high flow sources managed by groups with ample resources.

For basic overview of the activated sludge process, refer to the Compendium, pp. 124–125. For more technical and in-depth information on extended aeration package plants refer to US EPA (2000g) Wastewater Technology Fact Sheet on Package Plants: http://water.epa.gov/infrastructure/septic/upload/package_plant.pdf.
For more technical and in-depth information on SBRs refer to US EPA (1999c) Wastewater Technology Fact Sheet on Sequencing Batch Reactors: http://water.epa.gov/scitech/wastetech/upload/2002_10_31_mtb_sbr_new.pdf.
For comprehensive overview of RBCs refer to this SSWM link (Spuhler, 2010b): http://www.sswm.info/category/implementation-tools/wastewater-treatment/hardware/semi-centralised-wastewater-treatments/r.

Keys for selecting and planning secondary treatment systems

- understand regulatory discharge levels for treated effluent for the specific project, and utilize technologies that achieve these levels;
- given that there are a wide variety of treatment options that might be suitable for a given application, do not be swayed by sales pitches (especially related to the purchase of expensive mechanized options) without performing due diligence in the technology selection process;
- understand the full life cycle costs of technologies being considered to include capital cost to install, recurring costs for O&M, availability and cost of replacement parts, and longevity of the system. Learn about this by talking with owners of nearby facilities using the technologies before any purchase is made, and check for references;
- review agreements with subcontractors to verify that construction will meet all local and national requirements, tanks will be leak tested, the system will be started up and tested prior to turnover, and a full O&M plan will be provided upon completion.

4.7 TERTIARY TREATMENT

Even after secondary treatment, pathogen levels may be too high for safe discharge or reuse. Further treatment for microbial reduction can be achieved through filtration and disinfection, described in the following sections.

4.7.1 Tertiary Filtration

In some cases, when BOD and pathogen levels (especially for protozoans) in secondary effluent are still much higher than those required for discharge, DWMS should use a tertiary filtration step. This can involve membrane filtration, although slow sand filtration (shown in Figure 4.41) is generally less expensive and produces excellent results.

Figure 4.41 Slow sand filter in construction, showing gravel and sand layering, near Tullamore, Ireland (Molloy, 2013).

For more information about how tertiary filtration, clarification and other additional steps are used, especially within treatment plants to counteract high Phosphorus levels, refer to this US EPA (2007) resource: http://www.epa.gov/region10/pdf/tmdl/awt_report.pdf.

4.7.2 Disinfection

There are three common disinfection methods that may be appropriate for DWMS, which are

- radiation using ultraviolet (UV) light;
- chemical disinfection using chlorine;
- physical disinfection using ozone.

Since ozonation is rarely used in developing country settings, this manual will focus on disinfection by UV light and chlorination.

Ultraviolet light is a natural component of sunlight and can also be generated through special light bulbs. The higher energies of the ultraviolet spectrum are effective at inactivating microorganisms. While artificially-generated UV light from special bulbs is commonly used for disinfecting wastewater, it is mostly useful for large systems with high flow rates.

For more technical and in-depth information on UV bulb disinfection refer to US EPA (1999d) Wastewater Technology Fact Sheet on Ultraviolet Disinfection: http://water.epa.gov/scitech/wastetech/upload/2002_06_28_mtb_uv.pdf.

However, natural UV disinfection is practiced frequently in developing countries using shallow ponds, called maturation, aerobic, or polishing ponds as referred to in Section 4.6.3.3. These ponds, which can be installed as stand-alone units apart from WSP series, are designed with depths shallow enough to allow sunlight to penetrate throughout and disinfect the total depth of the water column. With appropriate loading from a secondary treatment system, and routine O&M to keep the surface free of floating and emergent vegetation, effluent can be sufficiently disinfected to meet discharge standards.

While polishing ponds have obvious advantages, the primary disinfection method used in water and wastewater treatment of all shapes and sizes worldwide is chlorination. Chlorine degrades pathogens upon contact for a short time, as long as it concentrated enough to avoid being entirely consumed by organic matter in the wastewater. Effective chlorination requires:

- proper chlorine concentrations;
- thorough mixing, to combine the dissolved chlorine with the wastewater;
- varying contact times, often between 20 minutes to 30 minutes depending upon ambient temperature.

The proper concentration of chlorine is added to the secondary effluent either as a diluted liquid that is dosed into the wastewater stream through pumps, or as a solid in the form of tablets that impart chlorine to the effluent as they dissolve. That concentration is based mostly on the how much organic matter is in the influent, as that can alter the composition of the chlorine and make it less effective for disinfection. This amount of chlorine that has been 'consumed' by the organic matter is called 'chlorine demand'.

While the feeder equipment costs are relatively low, the costs for the chlorine solution and O&M are considerably higher. The solid chlorine tablets are commercially available and are loaded into the stack delivery tube of the feeder (Figure 4.42). Routine cleaning is required to remove caked material in the tubes that can restrict complete mixing.

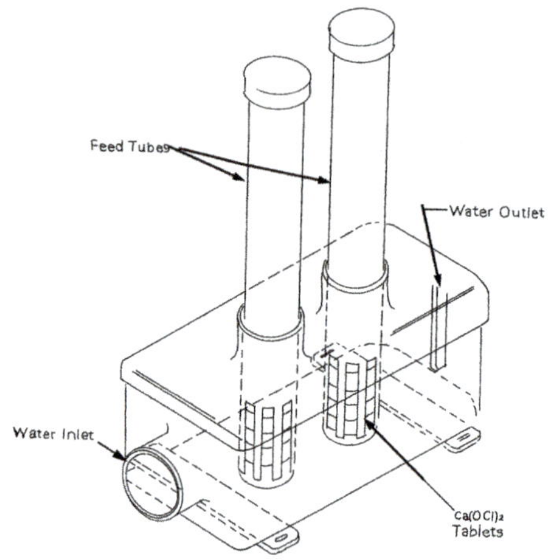

Figure 4.42 Stack chlorinator schematic (Jones, 2004).

The liquid chlorinators described above utilize liquid chlorine, typically in the form of sodium hypochlorite, which is liquid bleach. The bleach is mixed with the water in a feed tank, and a small metering pump is used to dose the effluent to the wastewater entering the contact basin. This is relatively simple technology but requires routine O&M.

For more technical and in-depth information on chlorination refer to US EPA (1999e) Wastewater Technology Fact Sheet on Chlorine Disinfection: http://water.epa.gov/scitech/wastetech/upload/2002_06_28_mtb_chlo.pdf.

Management considerations for disinfection systems

- certain pathogens (mainly protozoans) in the influent can be resistant to low doses of chlorine. Therefore, it might be necessary to opt for UV disinfection if the wastewater source is known to have high protozoan levels or if management of sludge (rich in protozoans) is perceived to be a weakness in the DWMS;
- if disinfection systems are mandated for a specific DWM application, monitoring the disinfection effectiveness through routine microbial laboratory tests is also likely needed. Laboratory testing is also useful because it can assess the secondary effluent chlorine demand, gives the initial chlorine concentration needed for dosage. Although rough estimates of chlorine demand for various wastewater sources can be found in the literature or by asking others for their experience, it is advisable to look into the laboratory testing route. This all suggests the need to verify local laboratory capabilities when assessing full life cycle costs;
- contacting chlorine, even in low dosages, can be hazardous for humans, and can be lethal to aquatic life in general. Wear eye and hand protection at all times when handling this corrosive and toxic substance, and ensure that the residual levels do not exceed discharge limits, especially if the receiving water body is a delicate and valued ecosystem.

4.8 END OF THE CYCLE – SAFE DISCHARGE OR REUSE

Discharge from DWMS refers to the drainage of the treated effluent off-site, into ditches or combined sewers, or directly into streams, rivers, estuaries, or oceans. Discharge from DWMS is permissible in most areas as long as the effluent complies with the limitations set by national, state or local governments. This generally entails permits that specify the type, number, and frequency of laboratory tests that must be run to verify that standards are met.

As mentioned in Chapters 2 and 3, reuse of treated effluent (often supplemented by biosolids usage) most frequently involves irrigation on landscape plants or food and non-food crops, and to a lesser extent, application for aquaculture, floating plant ponds, or toilet-flushing. Follow best practices to prevent cross contamination when dealing with reuse water plumbing inside of buildings, and over irrigation which may result in effluent runoff.

Safe reuse of wastewater for agriculture

In rural and peri-urban areas of most developing countries, the reuse of wastewater for irrigation is a common practice. Wastewater is often the only source of water for irrigation in these areas. Even in areas where other water sources exist, small-plot farmers often prefer partially or untreated wastewater because its high nutrient content reduces or even eliminates the need for expensive chemical fertilizers. Given this practice, the health and environmental risks do need to be carefully considered, but at the same time the importance of such reuse for the livelihoods of countless smallholders must also be considered.

These principles and safe reuse methods are the subject of a training handbook for farmer field schools published by the Food and Agriculture Organization of the United Nations, funded by the CGIAR Challenge Program on Water and Food. Titled 'On farm practices for safe use of wastewater in urban and peri-urban horticulture', it offers easy-to-understand explanations and useful practice exercises.

Download the guidebook (Beernaerts, 2012) and investigate the CGIAR program here: http://cgspace.cgiar.org/bitstream/handle/10568/21791/FFS%20Handbook_low.pdf?sequence=1.

The other major reuse processes include aquaculture and floating plant ponds. Aquaculture is the harvesting of fish that consume algae and other organisms within nutrient-rich water enhanced by the addition of the treated wastewater. The fish ponds require warm, tropical climates, and the fish must be able to tolerate water with low dissolved oxygen. While the fish do not greatly improve water quality, their economic and nutritional value (either as human or livestock feedstock) can offset system capital costs and provide a cheap source of protein.

The addition of effluent into floating plant ponds can be used to grow biomass in the form of water hyacinth, protein-rich duckweed and other macrophytes whose roots take up nutrients and further purify the effluent. These ponds are typically shallow and often utilize mechanical aeration, as their primary purpose is BOD and solids removal, while producing biomass. The biomass may have local economic value for use as animal feed, compost (as shown in the case study below), or green waste for processes such as anaerobic digestion. Treated effluent may also be used for interior use in buildings for toilet flushing in some areas, but strict controls on effluent quality and plumbing standards are required. In such situations, reuse plumbing is color-coded, and devices that protect potable water plumbing from cross-connections are mandated. As the price for potable water increases and its availability decreases, more innovative methods of effluent reuse and recycling will follow.

4.9 DETERMINING THE LEVEL OF REQUIRED TREATMENT

Determining the level of wastewater treatment required to meet specific discharge or reuse standards is an important step in the technology selection for DWMS. Making this determination in the planning stages of project development is critical. Such standards are necessary to ensure that the surrounding environment, particularly nearby waters intended for full-body contact or as drinking water supplies, are not adversely impacted by the discharge, and that aquatic life and human health are protected in reuse and other cases. Failure to provide enough treatment can result in notices of violation, fines, compliance orders, or mandatory retrofitting of the DWMS with additional treatment capacity, which implies the need to check with regulatory officials to fully understand discharge requirements for specific projects. On the other hand, providing more treatment than is needed may result in unnecessary capital and operational expenditures.

Co-managed wastewater resource recovery in the water-scarce province of Punata, Bolivia

Farmers in Punata reuse 100% of treated domestic wastewater to irrigate fodder crops during the driest half of the year. The wastewater treatment and resource recovery system consists of three AnPs in parallel, two FPs in parallel, three MPs in parallel, and a network of open channels to divert treated water to farms. It is co-managed by the municipal water company (AMAPAS) and a local farmers cooperative, which share different O&M responsibilities. For example, farmers trim vegetation around the lagoons and maintain irrigation canals (farmers discussing O&M of a nearby canal with researchers in Figure 4.43a), while AMAPAS is responsible for desludging the lagoons and maintaining the sewer collection system. The farmers recover nutrients from the wastewater by harvesting floating aquatic plants from the maturation ponds (Figure 4.43b shows pile harvested from pond at left). These are used either as animal feed or as compost, which after production in large piles is then mixed with topsoil to increase soil fertility.

Figure 4.43 (a) One farmer discusses resource recovery and O&M with a research group from the Universidad Mayor de San Simon (Verbyla, 2013); (b) Pile of algae dried after collection from maturation pond at left, with algae in maturation pond soon to be collected at right, in Punata, Bolivia (Verbyla, 2013).

However, flow design is not perfect, leading to short-circuiting and limited treatment efficiency (Mercado *et al.* 2013), with <90% helminth egg removal and <99% protozoan removal (Iriarte *et al.* 2013). While such co-management pond systems and reuse practices have great potential for small-scale application in water-scarce regions, more research and better design is needed to successfully scale up the treatment and irrigation systems to the size seen in Punata.

4.9.1 Meeting Discharge and Reuse Standards

Standards for discharging treated wastewater into the environment are highly variable and depend upon national standards, and in many cases, local requirements for specific watersheds. Often, the discharge standards are determined by the quality of the surface water where the discharge is intended. In The Philippines Class A waters are pristine and highly protected, requiring a very high degree of wastewater treatment. Class C rivers, typical in urban areas on the other hand, are already highly impacted by water pollution, and permit higher discharge concentrations of organic matter, suspended solids and pathogens.

In many developing countries, regulatory effluent standards are set at the national level for organic matter, suspended solids, various chemical compounds or nutrients, and FOG. An example for India is shown in Table 4.1, which lists those national standards based on the type of wastewater discharge (or irrigation reuse) as taken from their Environment Protection Act, Second Amendment Rules. Also, it is useful to consider how *E. coli* concentrations can be used to set reuse standards. This is demonstrated by the below reclaimed water classification system (as with Table 4.1, this is displayed not as endorsement but simply for use as background information for planner's reference in their own research) from the state of South Australia (Bolan *et al.*, 2008):

- **Class A** = less than 10 *E. Coli*/100 ml. Permitted uses include unrestricted irrigation of agricultural crops, such as spray or flood irrigation of surface crops and any irrigation of root crops;
- **Class B** = less than 100 *E. Coli*/100 ml. Permitted uses include drip or furrow irrigation of surface crops, subsurface irrigation of root crops, and spray irrigation of some crops with harvest controls; and
- **Class C** = less than 1,000 *E. Coli*/100 ml. Permitted uses include subsurface irrigation of surface crops, drip or furrow irrigation of crops with no ground contact and flooding or crops skinned before consumption with harvest controls.

Table 4.1 Government of India 1993 discharge standards for all wastewater (adapted from Reddy, 2010).

Parameter (mg/L)	Inland surface water	Public sewer	Land for irrigation	Marine coastal areas
Suspended solids	100	600	200	100
Oil and grease	10	20	10	20
Residual chlorine	1	n/a	n/a	1
Ammonia Nitrogen	50	50	n/a	50
Total Kjeldahl Nitrogen	100	n/a	n/a	100
Free Ammonia (NH3)	5	n/a	n/a	5
BOD	30	350	100	100
COD	250	250	250	250

Helminth and virus concentrations are also the subject of reuse guidelines, as promoted by the World Health Organization (WHO). To meet their health-based targets, wastewater reused in agriculture should have a mean value of less than one helminth egg per liter for 90 percent of all samples, while based on irrigation and crop conditions, there must be 99 percent to 99.99999 percent reduction in rotavirus concentration for safe reuse.

The WHO volumes on wastewater use in agriculture (Volume 2), wastewater and excreta use in aquaculture (Volume 3), and excreta and graywater use in agriculture (Volume 4) should be consulted prior to engaging in reuse. For further description, read the WHO (2006) Guidelines for the Safe Use of Wastewater, Excreta and Graywater: http://www.who.int/water_sanitation_health/wastewater/gsuww/en/.

Practical applications for effluent reuse

The ability to reuse treated effluent may be a significant driver for increasing willingness-to-pay for wastewater treatment. Follow this practical guidance to maximize that added value:

- Wastewater treatment to reuse standards may be expensive. Consider performing cost-benefit analysis for clarifying the proper level of treatment based on the costs and benefits of desired reuse activity and necessary additional treatment. That can constitute usage of additional DWMS technologies or coordinated practices such as drip or spray irrigation, or produce washing, disinfection, peeling or cooking (which have various microbe percent reduction estimates, as per the WHO Guidelines) that should be well-understood when planning reuse for irrigation;
- Consider how flow variability might impact reuse activities. For critical needs, such as agricultural irrigation, verify that alternative flow sources exist during times of low wastewater flow; and
- O&M of treatment systems for reuse, including filtration and disinfection systems, is critical to maintaining environmental health and safety. Determine special O&M requirements for reuse equipment and detail them in the O&M plan.

REFERENCES

Beernaerts I. (2012). On Farm Practices for Safe Use of Wastewater in Urban and Peri-urban Horticulture. http://cgspace.cgiar.org/bitstream/handle/10568/21791/FFS%20Handbook_low.pdf?sequence=1 (accessed 6 December 2013)

Bolan N., Laurenson S., Kunhikrishnan A., Naidu R., McKay J. and Keremane G. (2008). Northern Adelaide Plains Recycling Scheme – Champion in the Management of Recycled Water for Sustainable Production and Environmental Protection. http://irrigation.org.au/documents/publications-resources/conference_papers_2008/200508_Stream3_Bolan1.pdf (accessed 17 July 2013)

Casey P. (1996). Alternative Sewers: A Good Option for Many Communities. http://www.nesc.wvu.edu/pdf/WW/publications/pipline/PL_FA96.pdf (accessed 11 August 2013)

Consortium of Institutes for Decentralized Wastewater Treatment (2009). CIDWT Decentralized Wastewater Treatment Glossary, 2nd edn, 2009. http://www.onsiteconsortium.org/Glossary2009.pdf (accessed 4 April 2013)

Foster V. (2001). Lower Costs with Higher Benefits - El Alto Bolivia/Pilot Project. http://www.susana.org/docs_ccbk/susana_download/2-569-foster-2001b-lower-costs-higher-benefits-wsp-en.pdf (accessed 17 March 2013)

Gauss P. (2008). Constructed Wetlands: A Promising Wastewater Treatment System for Small Localities. http://www.wsp.org/sites/wsp.org/files/publications/ConstructedWetlands.pdf (accessed 12 November 2013)

Iriarte M., Mercado A., Verbyla M. E., Fuentes G., Rocha J. P. and Almanza M. (2013). Monitoring Indicators and Pathogenic Microorganisms into Two Stabilization Pond Systems in the Upper Valley of Cochabamba. Proceedings from the XV Congreso Bolivariano de Ingeniería Sanitaria y Medio Ambiente, ABIS-AIDIS, November 20–22, 2013, Cochabamba, Bolivia.

Jones D. (2004) Figure adapted by "The Disinfection Question – Answers for Onsite Systems", National Environmental Services Center. http://www.nesc.wvu.edu/pdf/WW/publications/pipline/PL_SP04.pdf (accessed 4 March 2013)

Jones D., Bauer J., Wise R. and Dunn A. (2001). Small Community Cluster Wastewater Systems. http://www.extension.purdue.edu/extmedia/ID/ID-265.pdf (accessed 12 February 2013)

Kayombo S., Mbwette T. S. A. and Katima J. H. Y. (2005). Waste Stabilization Ponds and Constructed Wetlands Design Manual. http://www.unep.org/ietc/Portals/136/Publications/Water&Sanitation/PondsAndWetlands_Design_Manual.pdf (accessed 27 July 2011).

Mara D., Lane J., Scott B. and Trouba D. (2010). Sanitation and Health. http://www.plosmedicine.org/article/info:doi/10.1371/journal.pmed.1000363 (accessed 12 November 2013)

Mercado A., Coronado O., Quinteros V. H., Verbyla M. E. and Oakley S. M. (2013). Evaluation of the Efficiency of the Wastewater Treatment Plant in Punata, Cochabamba, Bolivia: The Importance of Operation and Maintenance. Proceedings from the XV Congreso Bolivariano de Ingeniería Sanitaria y Medio Ambiente, ABIS-AIDIS, November 20–22, 2013, Cochabamba, Bolivia.

Molloy Precast Products, Ltd. (2013). Percolation and Tertiary Treatment. http://molloyprecast.com/percolation/tertiary.html (accessed 29 November 2013)

Parker, C. (2011). Cocopeat Biofiltration Systems. http://www.iwawaterwiki.org/xwiki/bin/view/Articles/CocopeatBiofiltrationSystems (accessed 12 April 2013)

Reddy K. L. D. (2010). TTI Consulting Engineers India, Ltd. Detailed Project Report: Design of Sewerage System in Kaveri Nagar of KR Puram CMC Area, Bangalore. Water and Sanitation for the Urban Poor-India, Bangalore.

Sasse L. (1998). DEWATS: Decentralized Wastewater Treatment in Developing Countries. http://www.sswm.info/sites/default/files/reference_attachments/SASSE%201998%20DEWATS%20Decentralised%20Wastewater%20Treatment%20in%20Developing%20Countries_0.pdf (accessed 5 October 2012).

Spuhler D. (2008). Anaerobic Filters. http://www.sswm.info/category/implementation-tools/wastewater-treatment/hardware/semi-centralised-wastewater-treatments-7 (accessed 25 May 2013)

Spuhler D. (2010a). UASB Reactor. http://www.sswm.info/category/implementation-tools/wastewater-treatment/hardware/semi-centralised-wastewater-treatments/u (accessed 25 May 2013)

Spuhler D. (2010b). Rotating Biological Contractors. http://www.sswm.info/category/implementation-tools/wastewater-treatment/hardware/semi-centralised-wastewater-treatments/r (accessed 25 May 2013)

Thermaco, Inc. (2013). About Us. http://www.trapzilla.com/about-us.php (accessed 2 December 2013)

UN-HABITAT (2008). Constructed Wetlands Manual. http://www.unhabitat.org/pmss/listItemDetails.aspx?publicationID=2983 (accessed 8 December 2012)

US EPA (1999a). Wastewater Technology Fact Sheet – Mound Systems. http://www.epa.gov/npdes/pubs/mound.pdf

US EPA (1999b). Wastewater Technology Fact Sheet – Intermittent Sand Filters. http://water.epa.gov/aboutow/owm/upload/2005_07_14_isf.pdf

US EPA (1999c). Wastewater technology fact sheet – Sequencing batch reactors. http://water.epa.gov/scitech/wastetech/upload/2002_10_31_mtb_sbr_new.pdf

US EPA (1999d). Wastewater Technology Fact Sheet – Ultraviolet Disinfection. http://water.epa.gov/scitech/wastetech/upload/2002_06_28_mtb_uv.pdf

US EPA (1999e). Wastewater Technology Fact Sheet – Chlorine Disinfection. http://water.epa.gov/scitech/wastetech/upload/2002_06_28_mtb_chlo.pdf

US EPA (2000a). Wastewater Technology Fact Sheet – Septic System Tank. http://water.epa.gov/scitech/wastetech/upload/2002_06_28_mtb_septic_system_tank.pdf

US EPA (2000b). Wastewater Technology Fact Sheet – Septic Tank Systems for Large Flow Applications. http://water.epa.gov/scitech/wastetech/upload/2002_06_28_mtb_septic_tank_large_flow_app.pdf

US EPA (2000c). Decentralized Systems Technology Fact Sheet – Septic Tank Leaching Chamber. http://water.epa.gov/scitech/wastetech/upload/2002_06_28_mtb_septic_tank_leaching_chamber.pdf

US EPA (2000d). Wastewater Technology Fact Sheet – Free Water Surface Wetlands. http://water.epa.gov/scitech/wastetech/upload/2002_06_28_mtb_free_water_surface_wetlands.pdf

US EPA (2000e). Wastewater Technology Fact Sheet – Subsurface Flow. http://water.epa.gov/scitech/wastetech/upload/2002_06_28_mtb_wetlands-subsurface_flow.pdf

US EPA (2000f). Wastewater Technology Fact Sheet – Trickling Filters. http://water.epa.gov/scitech/wastetech/upload/2002_06_28_mtb_ trickling_filter.pdf

US EPA (2000g). Wastewater Technology Fact Sheet – Package Plants. http://water.epa.gov/infrastructure/septic/upload/package_plant.pdf

US EPA (2002a). Wastewater Technology Fact Sheet, Sewers, Pressure. http://water.epa.gov/scitech/wastetech/upload/2002_10_15_mtb_presewer.pdf (accessed 5 November 2012)

US EPA (2002b). Wastewater Technology Fact Sheet – Anaerobic Lagoons. http://water.epa.gov/scitech/wastetech/upload/2002_10_15_mtb_alagoons.pdf

US EPA (2002c). Wastewater Technology Fact Sheet – Facultative Lagoons. http://water.epa.gov/scitech/wastetech/upload/2002_10_15_mtb_faclagon.pdf

US EPA (2002d). Wastewater Technology Fact Sheet – Aerated, Partial Mix Lagoons. http://www.epa.gov/owm/septic/pubs/apartlag.pdf

US EPA (2007). Advanced Wastewater Treatment to Achieve Low Concentration of Phosphorus. http://www.epa.gov/region10/pdf/tmdl/awt_report.pdf (accessed 12 Novemver 2013)

Verbyla M. (2013). Recent publications on the use of stabilization ponds in wastewater treatment and resource recovery systems. http://usf-reclaim.org/2013/12/stabilizationponds/ (accessed 13 December 2013)

Vymazal J. and Kröpfelová K. (2009). Removal of organics in constructed wetlands with horizontal sub-surface flow: a review of the field experience. *Science of the Total Environment,* **407**, 3911–3922. http://www.jlakes.org/web/Removal-organic-constructed-wetlands-STE2009.pdf

Wanasen S. A. (2003). Upgrading Conventional Septic Tanks by Integrating In-Tank Baffles. http://www.sswm.info/sites/default/files/reference_attachments/SRI%20ANANT%20W%202003%20Upgrading%20Conventional%20Septic%20Tanks%20by%20Integrating%20In%20Tank%20Baffles.pdf (accessed 7 October 2013)

Winblad U. and Simpson-Hebert M. (2004). Ecological Sanitation – Revised and Enlarged Edition. Stockholm Environment Institute, Stockholm, SE. ISBN: 91-88714-98-5.

World Health Organization (2006). Guidelines for the Safe Use of Wastewater, Excreta and Graywater, Vol **2–4**. http://www.who.int/water_sanitation_health/wastewater/gsuww/en/ (accessed 11 August 2013)

Chapter 5

Selecting wastewater technologies

This chapter will

- Discuss how selection of DWMS technologies can be informed by understanding the characteristics of a project's source and site and the available land area for managing the wastewater;
- Compare advantages and disadvantages of different DWMS technologies; and
- Help practitioners select the best technologies for their projects.

Figure 5.1 Sequencing Batch Reactor wastewater treatment package plant for public market and commercial center in San Fernando City, La Union, The Philippines. It is one example of how limited land availability can be the key driver in the DWMS technology decision-making process (Authors).

About Figure 5.1

Figure 5.1 is an example of how a highly mechanized technology might be appropriate for developing country settings. The goal of the project was to improve sanitation in the central business district in San Fernando City in La Union Province on the island of Luzon in the northern Philippines. San Fernando City does not have a centralized wastewater system, so DWMS are required to address wastewater discharges. For this project, wastewater from a number of commercial sources, including the public market, is collected in pump tanks that send the wastewater to the treatment plant. For this project, lack of space, high flow volume and high wastewater strength were the limiting factors.

To address this complex problem, system designers selected the sequencing batch reactor (SBR) with chlorine disinfection technology. SBRs, described in Chapter 4, are a form of aerobic treatment, but unlike their activated sludge counterparts, they require very little O&M from city maintenance personnel.

Among development experts, there is a debate about the merits of using mechanized technologies such as the controller-based SBR for wastewater projects. Many people believe that they are too complex and cities will not maintain them, and that electricity is too expensive and systems will be shut off to save power. While this may certainly be true of some areas, the real question is who will make this determination. For the San Fernando project, it was the city with support from the development team that decided upon this technology. The decision was based on the local capacity of the city engineering department and its staff to properly operate the system, the level of interest in improving sanitary conditions in the city expressed by the local and regional governments, and the support from stakeholders who were involved in the project from the outset.

In this case, now five years after its installation, it appears this was a sound decision, as the system is still functioning properly and has lived up to its expectations. San Fernando City provides a good example of the bottom-up planning process at work and how DWMS technology decision making can and should be a stakeholder-driven process. It is interesting to note that the early success of the public market project in San Fernando City has led to replication of the process for new wastewater systems for their slaughterhouse, a community interceptor system, and a newer comprehensive program for septage management. San Fernando City is well on its way to adopting DWMS as a viable model for wide-scale sanitation improvement throughout its urban areas. For more information about San Fernando City's wastewater interventions, visit its sanitation webpage at www.sfluseptage.blogspot.com.

5.1 INTERPRETING THE DATA COLLECTED FROM THE SOURCE AND SITE

Now that the source has been characterized and the site evaluated, the designer can begin to make some decisions regarding technology selection for the wastewater management system. That starts by assessing the data collected and determining the limiting conditions and specific opportunities presented by the project. The data collected up to this point can be summarized in the checklist represented as Table 5.1.

5.1.1 Flow to Land Availability Ratio Concept

Determining the land area that is available for the given volume of wastewater to be treated is a critical input for identifying appropriate technologies. After that is determined, the Flow to Land Availability Ratio (F:L) may be used to indicate which types of DWMS technologies might or might not be appropriate for the source and site. The F:L ratio is a number that is expressed in liters per day per square meter of available land:

$$\text{F:L} = \text{Design flow (L/day)}/\text{Land area available (square meters)}$$

Different wastewater management technologies have different space requirements. Sewage lagoons for example require ample land area, whereas aerobic treatment systems for the same flow have much lower space requirements. The F:L ratio helps quantify this concept.

In general terms, the higher the F:L ratio (more flow per square meter of available area), the more intensive the wastewater management needs to be. Continuing with the example, sewage lagoons and constructed wetlands are usually passive technologies requiring relatively little in the way of management, other than some periodic maintenance. Aerobic systems on the other hand, are highly mechanized and require significant O&M and electricity. These two technologies represent both ends of the spectrum of land requirement.

Refer to Table 5.2 where technologies are roughly ranked according to F:L ratio. Interpreting the table is simple; if the F:L for a source and site is above the recommended F:L range for a certain DWMS technology, that generally means that there is

Table 5.1 Summary checklist needed for technology consideration.

Parameters	Description	Yes	No
Flow rate	Has the design wastewater flow rate been determined?		
Flow strength	Has the organic loading rate been determined?		
Flow variability	Do flows vary daily, weekly or seasonally? Will flow equalization be useful?		
Nutrients and other considerations	Will special nutrient, grease, lint, trash, temperature, pH or other management be required?		
Site(s) selection and area(s)	Upon sketching the parcel, have the land surface areas for DWMS management been estimated?		
Soil horizons' characteristics	Have the soils been evaluated for depth, basic properties, and seasonally high groundwater/other limiting factors?		
Site slope and topography	Has the nature of the site's slope and landscape been identified and recorded?		
Site LTAR and other soils concerns	If soil dispersal is an option for the effluent, has an LTAR been estimated? If subsurface structures will be installed, has the stability of the soils been assessed?		
Surface waters	Have surface waters and related flood zones been identified and addressed?		
Land use	Have surrounding land use and potential conflicts been identified and addressed?		
Utilities and pre-existing DWMS	Are electricity and/or water available at the chosen locations, and have other infrastructure concerns such as pre-existing DWMS been identified and addressed?		
Access	Is access to the site for installation, O&M and other equipment sufficient and secure?		
Reuse and regulatory aspects	Have reuse opportunities and potential limiting regulations been identified and addressed?		

insufficient space on the parcel for that technology, and that another one with a greater F:L range should be chosen. While there are of course many other non-flow rate and non-land-area related factors that go into technology selection, which means that these F:L ranges can often be very flexible, the three ranges shown in Table 5.2, paired with their appropriate technologies, serve as good initial selection guidance.

Technology Selection for Wastewater Management at LORMA Medical Center:

Based on the source characterization and site evaluation, the management at LORMA knew the following:

- They had a small site and large flow (high F:L) that needed a highly mechanized DWMS;
- Due to the potential for floods, an above-ground system would be required;
- Reuse of the treated effluent was not a consideration, as management favored a discharging system;
- A gravity sewer could be used to connect the existing septic tanks, but a lift station would be required to pump the wastewater up to the treatment plant;
- Grease management was required due to the food preparation activities, but space for an outside grease interceptor was limited, so under-the-sink grease traps would be used;
- Lint management was required do to on-site laundry washing, and individual lint traps for each machine were chosen as that pre-treatment method.

An evaluation of the system four years after it was installed indicates that it is working well (Figure 5.2). The average O&M cost is approximately $175 per month including the operators' salary, electricity to run the system, and supply of treatment chemicals. This is a highly successful solution that indicates that even in developing country settings, there are applications for mechanized wastewater management systems.

Figure 5.2 The LORMA Medical Center SBR system installed by BioSafe, Inc. The cost was P 2 million ($46,000 USD), which was paid for by the facility owners (Authors).

For the LORMA project, management contacted a number of service providers and received quotations from several before selecting BioSafe, Inc. From this experience, LORMA President, Dr. Rufino Macagba offers the following lessons learned:

- the decision-makers should do their homework, starting with identifying at least three contractors using recommendations from other owners or operators of similar facilities that have already implemented systems;
- before selecting a bid, check references and verify that the proprietor is licensed to operate in the area and is in good standing;
- visit similar installations by the contractor and talk with owners about their experiences. This will help the implementer get a sense of what the system will look like once it is installed;
- ask for a construction schedule along with project milestones. Base payment on completion of milestones;
- ensure that the contactor is working with local regulators to secure all required permits and that all fees are paid;
- have a contract that specifies the services contracted for and the timeframe for delivery.

Table 5.2 F:L ratio and how it relates to technology selection.

F:L Ratio Range	Level of Mechanization	Appropriate Technologies – Primary Treatment	Appropriate Technologies – Secondary Treatment
Generally less than 100	Little to no mechanization	Anaerobic ponds	• soil dispersal systems in clayey soils; • drip irrigation systems; • facultative ponds; • aerobic ponds; • constructed wetlands
Generally between 100 and 500	Some mechanization, such as pumping systems	Anaerobic digester	• soil dispersal systems in sandy soils; • mound systems; • aerated ponds; • media filters
Generally greater than 500	Often full mechanization	Septic tanks, ABRs	• activated sludge; • SBRs; • RBCs; • extended aeration

With this information, and that from Table 5.1, the practitioner has enough information to start identifying which technologies might be appropriate for their specific project. The following sections will address how potentially appropriate technologies can be further narrowed down to the most suitable ones given the specific source flow and site characteristics of the project, as well as the technologies' various resource requirements.

5.2 TECHNOLOGY SELECTION STRATEGIES

The purpose of the following sections is to organize the information about each of the technologies discussed so far in this manual, so as to give additional guidance for the selection process. The guidance is not intended to be exhaustive, as there are dozens of wastewater technologies that might be used for any given wastewater management need, and new technologies are being developed each year. Most fall within the categories presented or are close enough to others so that the reader can make comparisons. More importantly, the information is provided so that readers can better understand the relationship between source characteristics and the site constraints or limiting conditions, and how considering both can lead towards effective DWMS technology selection.

Tables 5.3 to 5.7 list various characteristics of the presented DWMS technologies grouped based on their order within the DWMS Developmental Model schematic, which groups technologies as follows:

- pretreatment technologies (Table 5.3);
- conveyance technologies (Table 5.4);
- primary treatment technologies (Table 5.5);
- secondary treatment technologies (Table 5.6);
- tertiary treatment technologies (Table 5.7).

The tables rate how each DWMS technology can be defined in terms of its source and site considerations as well as cost, O&M and treatment effectiveness. The tables simply intend to help compare general technologies and are not to be considered as full recommendations for the selection of any technologies for a given project. In general, readers are encouraged to use the information in the tables to make a general assessment of which technologies might be appropriate for their given needs, and to then further research technologies prior to making final selections. Sources of additional information may include the references provided in this manual, contractors that are providing wastewater system services, users of wastewater technologies, local government officials, and civil society groups that may be locally engaged in this field.

For more in-depth comparison of most of the primary and secondary treatment technologies described in the following tables, refer to pp 234 and other sections of Marcos von Sperling's publication on biological wastewater treatment: http://www.iwawaterwiki.org/xwiki/bin/download/Articles/DevelopingCountriesTitlesfromIWAPublishingFreetoDownload/Wastwater.pdf.

Table 5.3 Pre treatment technology comparison guide, based on appropriate levels for the various key parameters ('any' is any level of the parameter is accepted, and 'N/A' means the parameter is not applicable for the technology).

Technology	Flow Rate	Flow Strength	Flow Variability Tolerated	Space Needs	Site Slope Tolerated	Need for Good, Deep Soils	Cost	O&M	Treatment effectiveness	Other concerns
Under the sink grease trap	Low	Medium	Low – can wash out during high flows	Low	Any	N/A	Low	High – should service daily	Medium	May generate odor in the kitchen when serviced
Outdoor grease interceptor	Medium to high	Any	Any	High – usually installed in parking areas	Low	Medium	Medium	Low – should pump the tank every few months	Medium	Must be traffic-rated if installed in parking area; collected grease is saturated and not useful as feedstock for commercial products
High efficiency grease interceptor	Any	High	Any	Medium – can be installed inside the kitchen or outside the facility	Any	Low	High	Medium – remove grease every month; can be done by hand or with pump truck	High	Product may not be locally available; collected grease is nearly pure and can be used as feedstock for biodiesel or other products
Trash trap	Any	Any	Any	Low	Any	N/A	Low	High – clean daily	Medium	Odor if not regularly maintained; sewer clogging if not cleaned often
Lint trap	Any	Any	Any	Low	N/A	N/A	Low	High – clean daily	Medium	Products adapted for use on individual machines, or entire system

Table 5.4 Sewerage technology comparison guide, based on appropriate levels for the various key parameters ('any' is any level of the parameter is accepted, and 'N/A' means the parameter is not applicable for the technology).

Technology	Flow Rate	Flow Strength	Flow Variability Tolerated	Land Needs	Site Slope Tolerated	Need for Good, Deep Soils	Cost	O&M	Treatment effectiveness	Other concerns
 Building sewers	Any	Any	Any – make sure to size for peak flows	Low	Any – 1% to 2% for gravity flow; use drop boxes for steep sites	Low – generally excavated at shallow depths	Low	Low	N/A	Keep large solids and trash out of the sewer; use sealed and locking access ports for safety
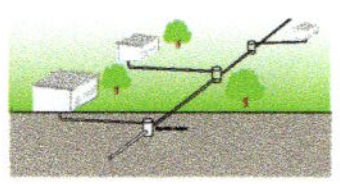 Simplified sewers	Low to medium	Low	Low to medium	Low	Low	Low to medium	Low	Medium- participation on part of homeowners is necessary	N/A	Dense residential land use preferrred
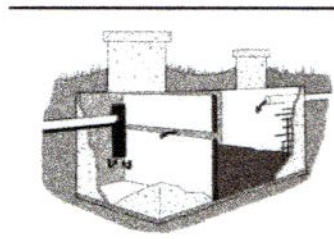 Solids-free sewers	Low to medium	Medium to high	Low to medium	Medium	Low to medium	Medium – high groundwater or bedrock can limit depth of tank excavation	Medium to high	Medium to high	Medium	Must have septage management program and frequent cleaning of tanks
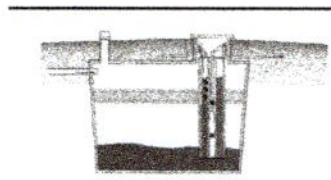 Pressure sewers	Low to medium	Low to medium	Low to medium	Medium	Any	Medium to high	High	High	High	Electricity needed for pumping

Table 5.5 Primary treatment technology comparison guide, based on appropriate levels for the various key parameters ('any' is any level of the parameter is accepted, and 'N/A' means the parameter is not applicable for the technology).

Technology	Flow Rate	Flow Strength	Flow Variability Tolerated	Land Needs	Site Slope Tolerated	Need for Good, Deep Soils	Cost	O&M	Treatment effectiveness	Other concerns
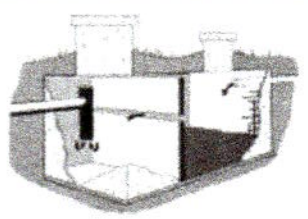 Septic tank	Low	Low to medium – usually for residential source(s)	Medium	Low	Medium to high	Medium – high groundwater or bedrock can limit depth of tank excavation	Medium	Low	Low	Odor if not properly vented
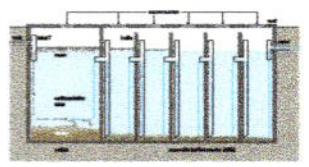 Anaerobic Baffled Reactor	Medium	Medium – for residential and/or commercial source(s)	Medium	Low to medium	Medium to high	Medium – high groundwater or bedrock can limit depth of tank excavation	Medium	Low	Low to medium	Odor if not properly vented
 Anaerobic Biodigester	Low to medium	High – animal manure, food waste best	Low	Medium	Low to medium	Depends on type – medium for domes and floating drums, low for tubular units	Medium to high	High	Medium	Odor; gas production monitoring and safety; residual biogas and biosolid usage
 Anaerobic Lagoon	High	High – for standard strong influent, rather than animal or other waste	High	High	Low	High – depth range of 2–5 m is often limited by groundwater or bedrock	Low – assuming cheap land	Low	Medium	Odor; mosquitos; maintaining proper distance from population centers

Table 5.6 Secondary treatment technology comparison guide, based on appropriate levels for the various key parameters ('any' is any level of the parameter is accepted, and 'N/A' means the parameter is not applicable for the technology).

Technology	Flow Rate	Flow Strength	Flow Variability Tolerated	Land Needs	Site Slope Tolerated	Need for Good, Deep Soils	Cost	O&M	Treatment effectiveness	Other concerns
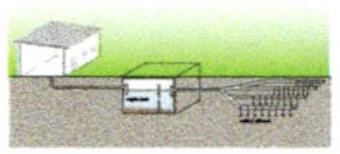 Non-mechanical soil dispersal systems	Low	Low	Any	Medium to high	Low to medium	Medium – clay soils are not ideal, and increase land needs	Low	Low	Medium to high	Improper soils evaluation can lead to failure or aquifer contamination
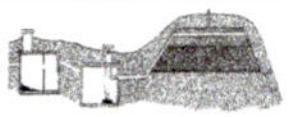 Mechanical soil dispersal systems (mounds)	Low	Low	Any	Medium	Low	Medium – clay soils may be limiting; shallow soils are OK	Low to medium	Medium	High	Electricity needed for pumping
 Mechanical soil dispersal systems (drip irrigation)	Low to medium	Low to medium	Any	Medium to high	Low to medium	Medium – clay soils may be limiting; shallow soils are OK	Low to medium	Low to medium – unclogging of pipe often necessary	High	Lack of locally-available materials may be limiting factor
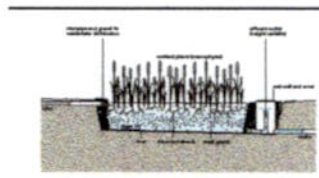 Constructed wetlands (horizontal)	Low to medium	Low to medium	Any	High	Low	Medium – usually requires excavation to 1 m depth; clay soils may not require liner	Medium	Low	Low to medium	Opportunity to merge wastewater treatment with habitat creation
 Facultative lagoon	Medium to high	Medium to high	Any	High	Low	High – generally must excavate to 2–3 m depth	Medium	Low – O&M still required to minimize nuisance issues	Low to medium	Odor; mosquitos
 Media filter	Low to medium	Low to medium	Any	Low	N/A	N/A	Medium	Medium – regular replacement of media	Medium to high	Requires locally-available media, and warm climate; sludge production
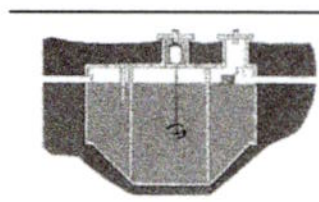 Aerobic system	High	Medium to high	Medium	Low	N/A	N/A	High	High – requires skilled construction and O&M staff	High	Noise; electricity demand; sludge production

Table 5.7 Tertiary treatment technology comparison guide, based on appropriate levels for the various key parameters ('any' is any level of the parameter is accepted, and 'N/A' means the parameter is not applicable for the technology).

Technology	Flow Rate	Flow Strength	Flow Variability Tolerated	Land Needs	Site Slope Tolerated	Need for Good, Deep Soils	Cost	O&M	Treatment effectiveness	Other concerns
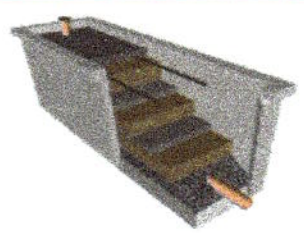 Sand filtration	Low to medium	Medium to high - ideal for strong secondary effluent	Low	Low	Any	Low	Medium – media may be expensive	Medium	High	Effective secondary treatment required prior to tertiary filtration
 Maturation pond	Any	Low	Any	High	Low	Medium	Medium – like with other lagoons and wetlands, depends on excavation costs	Low to medium	Low	algae production; mosquitos
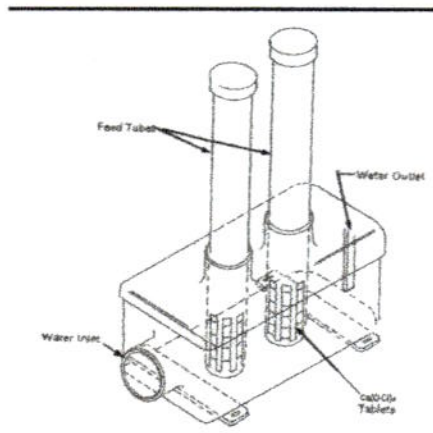 Chlorination	Any	Low – often requires tertiary filtration beforehand	Low to medium	Low	Any	N/A	Medium – device and chemical solution may be costly	Medium	Medium	Electricity needed for dosing; safety in handling

5.2.1 Additional Considerations for Technology Selection

The key to effective wastewater pretreatment is O&M, with grease traps being a good example. The under-the-sink grease trap is a low-cost system that will work well if properly maintained. The outdoor grease trap is much easier to maintain, but costs more to pump periodically, whereas the high-efficiency grease interceptor is more expensive, but produces a usable byproduct.

Carefully planning the routes of sewers to minimize slope variability helps keep costs down, while failure can be avoided by ensuring that pipes do not leak and are properly protected with bedding material under the pipe and enough soil cover above the pipe. Outside of proper installation, user education and preventative O&M are keys to good sewer operation, while for tank-using sewers, proper pre-treatment (locking access ports to keep out large solids and trash) and comprehensive septage management must also be implemented.

When all of these conditions are met, innovative sewerage methods like solids-free, simplified and pressure sewer systems can be successfully applied in developing country settings. As one example, Panglao Island, a resort community in The Philippines, utilizes pump systems for each of its beachside hotels to remove the sewage from the waterfront and pump it up to a central location for treatment. Also, simplified sewerage has been used with great success in Brazil and other locations for communities of more than 20,000 people.

Primary treatment systems remove the heavy solids in wastewater through settling, and the lighter fats, oil and grease through flotation. Septic tanks can work well when properly sized and maintained. ABRs are more efficient than septic tanks but still require secondary treatment. Anaerobic digestion for biogas (methane) production is viable if the waste contains a high percentage of volatile solids, as with fresh animal manure. Such waste may be mixed with properly pre-treated residential wastewater, crop waste or food waste to improve methane production. Use skilled engineering to best design systems even when using the toolkit in Section 4.5.3. Anaerobic sewage lagoons are recommended provided they can be located well away from population centers. Properly designed systems will provide a moderate degree of wastewater treatment, often to secondary standards, although seasonal variability might impact on effluent quality. Facultative or aerobic ponds, or constructed wetlands, are often installed after these lagoons.

Selecting appropriate secondary treatment technologies is a function of:

- the amount of land available for the system (F:L ratio);
- the adequacy of the soils for soils-based dispersal;
- the proximity of the project site to population centers;
- the skill level of the workforce that will perform system O&M; and
- the funds available for installation, O&M and eventual replacement.

Beyond these factors, a careful assessment of the social, financial and governance aspects of the community or project that will be served by the system is required. For larger projects, governance (the ability of the local government or utility district to charge and collect fees, and levy fines and penalties for non-compliance) has a significant impact on DWMS technology selection. If funds will be limited to system O&M once it is built, this becomes a key driver in the decision-making process. Carefully considering all of these factors, and doing so using an open, transparent and stakeholder-driven process, can lead to better technology choices and a greater willingness-to-pay for services over the long-term.

Wastewater effluent that will be reused in buildings for toilet flushing, discharged into water bodies, or recycled for use as crop irrigation water usually needs to receive tertiary treatment, which is filtration and disinfection. The filtration step, critical for effective pathogen removal, is reliably done through sand filtration. While this technology has not been presented at length in this manual, there is a lot of good information easily accessible from the references section of Chapter 4.

Sand filtration removes most of the remaining suspended solids in the secondary effluent discharge. This is important as suspended solids can shield pathogens from the effects of chlorine or UV radiation. Once influent is filtered, low levels of chlorine or UV light can dramatically reduce the number of pathogens in the wastewater.

If land is available, consider using maturation ponds for pathogen control. If land is limited, chlorine may be the only other practical disinfection option. Carefully consider the benefits and detriments of both liquid and solid chlorine, their safety concerns, and long-term costs and product availability prior to making a technology selection.

5.3 CONCLUSION AND LOOKING FORWARD

This manual presents the step-by-step procedures for characterizing wastewater sources, evaluating sites where wastewater will be managed, and using the data collected and DWMS technology parameters to help inform DWMS technology-selection decisions. The guidance has been presented from the standpoint of the *individual project*, such as for a school, hospital or housing block. As more and more individual projects are developed in a given community, and the concepts of DWMS takes hold, cities and municipalities will begin to promote DWMS as a means for affecting sanitation improvements on a large scale. Local governments can do a lot to achieve this proactively through the following steps:

- Promote the benefits of improved sanitation by first highlighting the problem and then describing how DWMS can be part of the solution. Utilize stakeholders, NGOs, and civil society groups to spread the word. As people begin to see the benefits of improved sanitation and understand how it will positively impact their lives and communities, they will be more willing to pay for services;
- Develop policies and approve local ordinances on sanitation that require proper wastewater management, especially for new construction, and for wastewater sources that are deemed to be imminent health hazards;
- Train staff on DWMS technology types and the step-by-step methods described in this manual. There are many free resources that can be used to begin the process of training staff. Work with local universities whenever possible. Utilize twinning and job-sharing with municipalities that are further ahead in the process as one way of building local capacity;
- Link the requirements for proper DWMS with issuance of building permits. Train building inspectors on the procedures for inspecting septic tanks and other DWMS technologies;
- Develop a 'one-stop-shop' style of permit acceptance and review, making it as easy as possible for citizens to comply. Offer technical assistance as necessary;
- Promote good construction practices for septic tanks and other DWMS technologies through workshops, seminars and hands on training exercises; and
- Ask for help. After many years of trying, the development community has come to realize that development works best when it is demand-driven, rather than when it is forced upon a community. Therefore, development groups are eager to assist those that come to them for help (demanding services). Use the power of the Internet to reach out to development groups, many of which are listed in the appendices of this manual's chapters.

The authors wish to thank you for your interest in this topic and look forward to your email comments.

REFERENCE

von Sperling M. (2013). Wastewater Characteristics, Treatment and Disposal: Volume One, Biological Wastewater Treatment Series. http://www.iwawaterwiki.org/xwiki/bin/download/Articles/DevelopingCountriesTitlesfromIWAPublishingFreetoDownload/Wastwater.pdf (accessed 29 November 2013).

Appendices

Practical application of soils evaluation data

Understanding the nature of soils on a given site is critical for proper implementation of DWMS, not only for basing technology decisions, but also for ensuring proper installation. Chapter 3 describes basic procedures for how service providers can conduct proper soils evaluations. The following pages provide supplementary information and illustrate how all findings can be organized in a standard format for the purposes of recording and interpreting the data. The appendices are given in the following order:

- *Appendix I* provides additional information and definitions of various soils properties and procedures used within soils evaluation, as well as a standard test hole soils evaluation log form, along with tables for adjusting the Long Term Acceptance Rate (LTAR) based on specific soils conditions that are identified, with the ultimate goal of determining the LTAR for the soils on a site;
- *Appendix II* provides an example LTAR calculation based on a hypothetical site, allowing readers to see how soils data is actually reported and used to determine the ultimate LTAR.

Many researchers have contributed to the practice of soils evaluations for DWMS. The authors would like to recognize the following individuals and organizations that have significantly contributed to this body of soils science and translated it into a practical approach that can be used anywhere to improve the long-term function of soils-based DWMS components:

David Lindbo
North Carolina State University,
Soil Science Department, P.O. Box 762,
Raleigh, NC 27695-7619 (david_lindbo@ncsu.edu)

Randy Miles
University of Missouri,
302 Anheuser-Busch Building,
Columbia, MO 65211

DeAnn Presley
Kansas State University,
2014 Throckmorton Plant Science Center,
Manhattan, KS 66506-5504

Special thanks to **The Soil Science Society of America**, www.soils.org for permission to reprint their resources, as well as David Lindbo for several images and tables, cited as (Lindbo, year produced). The main resources used here, referred to as (Lindbo *et al.*, 2008) and (Lindbo, 2009) have the following citations:

Lindbo D., Miles R. and Pressley D. (2008) Soil Profile Descriptions. *In: Soil Science: Step-by-Step Field Analysis.* S. Logsdon, D. Clay, D. Moore and T. Tsegaye Soil Science Society of America, Madison, WI, U.S., 11–34.

Lindbo D. (2009) Chapter 4: Soil and Site Concepts for Installers. Installation of Wastewater Treatment Systems. Consortium of Institutes for Decentralized Wastewater Treatment (CIDWT), Iowa State University, Midwest Plan Service. Ames, IA. pp. 43–62.

Appendix I

More on soils evaluation

A1.1 OVERVIEW OF THE SOILS EVALUATION PROCESS

The soils evaluation procedures presented in Section 3.3 introduce the concept of using simple tests that can be performed in the field to learn about key aspects of the soils. The depth and thickness of the soils horizons and the presence of groundwater or bedrock should be determined during all DWMS projects, as they may have an impact upon the installation of tanks, trenches and basins. The other features such as soil texture, structure, consistence, as well as slope description and landscape position also determine how well the soils will perform for both dispersal and treatment. Each of these factors is presented in the following pages that provide a step-by-step procedure for evaluating the soils for on-site effluent dispersal.

While the main goal of these exercises is to determine the ultimate Long Term Acceptance Rate (LTAR) of each soils horizon of significance at a particular site, it is interesting to note that some of these parameters also may have a bearing on the system installation. As one example, the degree of soil 'stickiness', one of the features of the soils evaluation procedure presented below, can have a profound impact on the installation of leach trenches. For soils that are moderate to very sticky, installation should be limited to the driest time of the year.

Before beginning the discussion on determining the LTAR, a note on the units used in these appendices is in order. Much of the work on and research into soil science has been conducted in the United States. Therefore, for convention, the units of measure are presented in 'English' measurements. A simple conversion formula for LTAR units in the U.S., in gallons per day per square foot (gpd/sf), is provided below:

To convert to metric: $\text{gpd/sf} \times 40.7 = \text{liters/day/square meter}$

This is an expression of how much volume (gallons) can be applied to an area of soil (square foot) each day. Area in this case is described as the infiltrative surface, or the area where the leaching structure meets the natural soil. For a leach trench, it is the side wall and bottom, although in certain areas and within this manual, the bottom is not counted to make design more conservative. Remember that for leach trenches, there are 2 sidewalls, and the area of each measured from below the distribution pipe is considered to be the infiltrative surface. For leach beds, the infiltrative surface is the bottom of the bed. For drip irrigation fields, the infiltrative surface is the entire area of the field.

The following tables (adapted from: D. Lindbo, personal communication, July 8, 2013) are utilized in the step-by-step process of determining LTAR, and will be referred to in the following pages based on how they give LTAR adjustment values for various soil property conditions. The tables are grouped here for convenience. Each is described in detail in the text sections that follow.

Table A1.1 Soil texture.

USDA Texture	Group/Range	Mid-point	LTAR Adjustment
Sand (S)	I 1.2–0.8	1	na
Loamy sand (LS)			na
Sandy loam (SL)	II 0.8–0.6	0.7	na
Loam (L)			na
Silty loam (SiL)	III 0.6–0.3	0.45	−0.05
Silt (Si)			−0.1
Sandy clay loam (SCL)			na
Clay loam (CL)			na
Silty clay loam (SiCL)			−0.05
Sandy clay (SC)	IV 0.4–0.1	0.25	na
Clay (C)			na
Silty clay (SiC)			−0.05

Table A1.2 Other texture considerations and their adjustments.

Mica	LTAR Adjustment	% Coarse Fragments	LTAR Adjustment
Paramicaeous	−0.05	0–15	na
Micaeous	−0.1	15–35	−0.05
		35–60	−0.1
		60+	UNS

Table A1.3 Soil structure – grade, size and type.

Grade	LTAR Adjustment	Size	LTAR Adjustment	Type	LTAR Adjustment
0 – SG – Single grain	na	vf – Very fine	0.05	Gr – Granular	0.05
0 – M – Massive	UNS	f – Fine	0.05	SBK – Subangular blocky	Na
0 – RCF – Rock controlled fabric	−0.05	m – Medium	*na*	ABK – Angular blocky	−0.05
1 – weak	−0.05	c – Coarse (to 1")	−0.05	W – Wedge	UNS
2 – moderate	Na	vc – Very coarse	UNS	PR – Prismatic	UNS
3 – strong	0.05			PL – Platy	UNS

Table A1.4 Soil consistence and mineralogy (see footnote).

Moist	LTAR Adjustment	Wet – Stickiness	LTAR Adjustment	Wet – Plasticity	LTAR Adjustment
Loose	na	NS	0.05	NP	0.05
vFr – Very friable	0.05	SS	na	SP	na
Fr – Friable	na	MS	−0.05	MP	−0.05
Fi – Firm	−0.05	VS	UNS	VP	UNS
vFi – Very firm	UNS				
H – Hard	UNS				

Make only 1 LTAR adjustment based on consistence tests, use most restrictive

Table A1.5 Soil depth to unsuitable conditions.

Depth beneath infiltrative surface in Group II–IV textures	Depth beneath infiltrative surface in Group I texture	LTAR adjustment
<12″	<18″	UNS
12–18″	18–24″	−0.05
18 to 24″	24 to 30″	na
>24″	>30″	0.05

Table A1.6 Landscape (slope) position.

Slope position	LTAR Adjustment
Summit or ridge	0.05
Shoulder	0.05
Back or side slope (upper)	na
Back or side slope (lower)	−0.05
Foot slope	−0.05
Toe slope	−0.1
Channel or Drainage way	UNS
Depression	UNS
Floodplain	−0.1

Table A1.7 Slope description.

Slope description	LTAR Adjustment
VV–Convex, convex	0.05
LV–Linear, convex	0.05
VL–Convex, linear	na
LL–Linear, linear	na
CV–Concave, convex	−0.05
LC–Linear, concave	−0.05
VC–Convex, concave	−0.1
CL–Concave, linear	−0.1
CC–Concave, concave	−0.15

Table A1.8 Slope correction, requiring addition of soil above down-slope trench wall bottom for trenches perpendicular to hill slope, of up to 22 inches depth based on slope and trench width.

% Slope	Trench width (inches)						% Slope	Trench width (inches)					
	36	30	24	18	12	6		36	30	24	18	12	6
5	2	2	1	1	1	0	30	11	9	7	5	4	2
6	2	2	1	1	1	0	31	11	9	7	6	4	2
7	3	2	2	1	1	0	32	12	10	8	6	4	2
8	3	2	2	1	1	0	33	12	10	8	6	4	2
9	3	3	2	2	1	1	34	12	10	8	6	4	2
10	4	3	2	2	1	1	35	13	11	8	6	4	2
11	4	3	3	2	1	1	36	13	11	9	6	4	2

(*Continued*)

Table A1.8 Slope correction, requiring addition of soil above down-slope trench wall bottom for trenches perpendicular to hill slope, of up to 22 inches depth based on slope and trench width (*Continued*).

% Slope	Trench width (inches)						% Slope	Trench width (inches)					
	36	30	24	18	12	6		36	30	24	18	12	6
12	4	4	3	2	1	1	37	13	11	9	7	4	2
13	5	4	3	2	2	1	38	14	11	9	7	5	2
14	5	4	3	3	2	1	39	14	12	9	7	5	2
15	5	5	4	3	2	1	40	14	12	10	7	5	2
16	6	5	4	3	2	1	41	15	12	10	7	5	2
17	6	5	4	3	2	1	42	15	13	10	8	5	3
18	6	5	4	3	2	1	43	15	13	10	8	5	3
19	7	6	5	3	2	1	44	16	13	11	8	5	3
20	7	6	5	4	2	1	45	16	14	11	8	5	3
21	8	6	5	4	3	1	46	17	14	11	8	6	3
22	8	7	5	4	3	1	47	17	14	11	8	6	3
23	8	7	6	4	3	1	48	17	14	12	9	6	3
24	9	7	6	4	3	1	49	18	15	12	9	6	3
25	9	8	6	5	3	2	50	18	15	12	9	6	3
26	9	8	6	5	3	2	51	18	15	12	9	6	3
27	10	8	6	5	3	2	52	19	16	12	9	6	3
28	10	8	7	5	3	2	53	19	16	13	10	6	3
29	10	9	7	5	3	2	54	19	16	13	10	6	3
30	11	9	7	5	4	2	55	20	17	13	10	7	3
							56	20	17	13	10	7	3
							57	21	17	14	10	7	3
							58	21	17	14	10	7	3
							59	21	18	14	11	7	4
							60	22	18	14	11	7	4

$$A = W \times (S/100)$$

A1.2 SOILS EVALUATION METHODS AND PROCEDURES

A1.2.1 Soil Texture and Structure

The procedures for conducting soil texture analysis were covered thoroughly in Section 3.3.3.2. The next step is to determine the soil structure, which is defined as soil peds' type, size, and grade. Ped type is very easy to discern, simply by comparing the ped with Figure A1.1. The primary six categories (used for planning dispersal systems) are granular, subgranular blocky, angular block, wedge, prismatic and platy. The other categories indicate poor soils (horizons) that should not be used for dispersal.

Understanding ped type informs the designation of ped size, which requires knowing the ped type (class) and then measuring its thickness or diameter (in general, its smallest dimension). The ped type determines which column of Table A1.9 must be scanned, while that measurement (as designated by the column headers for the grouped ped classes) determines which row should be selected. Each row, when followed to the left, reveals the proper size classification of the ped. It is important to know what to measure for each class type (thickness or diameter) and to keep in mind that having very coarse (or thicker) size peds in a horizon may make it unsuitable for dispersal systems.

Ped grade can be broken down into four groups as follows:

- *Structureless*: *Group 0*: no distinct peds can be viewed in the horizon or a hand sample;
- *Weak*: *Group 1*: the peds can barely be seen as distinct in the horizon or hand sample;
- *Moderate*: *Group 2*: the peds are clearly seen when viewed in the horizon and hand sample; and
- *Strong*: *Group 3*: the peds are very distinct-looking and separate easily from each other when moved.

Given how each of the three components of soil structure influences the flow of effluent through soil, their consideration is required in determining the ultimate LTAR. Refer to Table A1.3 for the appropriate adjustment factors related to soil structure.

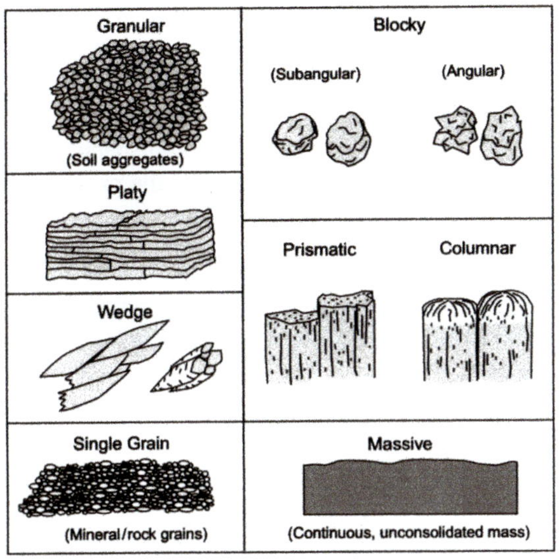

Figure A1.1 Common soil structure ped types (Lindbo *et al.*, 2008).

Table A1.9 Ped size classification based on ped type and dimensions (adapted from Lindbo *et al.*, 2008).

Class	Granule or plate thickness size**	Column, prism, or wedge diameter size**	Angular or subangular blocky size**
Very fine or very thin*	<1	<10	<5
Fine or thin*	1 to <2	10 to <20	5 to <10
Medium	2 to <5	20 to <50	10 to <20
Coarse or thick	5 to <10	50 to <100	20 to <50
Very coarse or very thick	10+	100 to <500	50+
Extremely coarse	–	500+	–

*Thin is only used for plates
**Size is for the smallest dimension in mm.

A1.2.2 Soil Consistence

Consistence is judged using these three major methods given below, which all factor into the LTAR calculation:

- *Resistance to breaking*: This labels the ped as having a moisture class (Table A1.10) based on how much force the (moist) ped can handle when being compressed between the thumb and index finger before it ruptures. To test this, take a ped block with edge of around 3 cm (or plate ped around 1 cm long and 0.5 cm thick) and push it between those fingers for one second. Start with very little force, and increase that each time until the ped breaks. When it does, the appropriate moisture class, of which there are five ranging between loose and very firm (which is unsuitable for dispersal), can be judged based on the table.

Table A1.10 Classifying moisture level based on rupture resistance (adapted from Lindbo *et al.*, 2008).

Moisture Class	Operation
Loose	Specimen not obtainable
Very friable	Very slight force between fingers
Friable	Slight force between fingers
Firm	Moderate force between fingers
Very firm	Strong force between fingers

- *Stickiness*: This refers to how much the ped sticks to other things. It is tested by crushing peds in one's hand and pushing the soil through the thumb and index finger while pouring water on the soil until it sticks as much as it possibly can to the fingers. Run this test a few times with different water-pouring levels to find out which moisture level gives the maximum adherence for the sample. The overall stickiness is judged at that moisture level, and classified at four levels between non-sticky (NS) and very sticky (VS, which is unsuitable for dispersal). This classification, done according to Table A1.11, is based on how much soil sticks to the fingers after the pressure is released.

Table A1.11 Stickiness classification system (adapted from Lindbo *et al.*, 2008).

Class	Stickiness Criteria
Non-sticky	Little or no soil adheres to fingers after release of pressure
Slightly sticky	Soil adheres to both fingers after release of pressure with little stretching on separation of fingers
Moderately sticky	Soil adheres to both fingers after release of pressure with some stretching on separation of fingers
Very sticky	Soil adheres firmly to both fingers after release of pressure with much stretching on separation of fingers

- *Plasticity*: This is how much soil can be twisted and turned as one piece until it breaks. Take soil at the same moisture level as was used for the stickiness test, and roll it between the hands into as long and thin of a cylinder as possible. The diameter and strength of the roll can be used to classify the soil sample into four categories of plasticity, ranging from non-plastic (NP) to very plastic (VP, unsuitable for dispersal), as shown in Table A1.12.

Table A1.12 Plasticity classification system (adapted from Lindbo *et al.*, 2008).

Class	Plasticity Criteria
Non-plastic	Will not form a 6 mm diameter roll, or if formed, cannot support itself if held on end
Slightly plastic	6 mm-diameter roll supports itself; 4 mm-diameter roll does not
Moderately plastic	4 mm-diameter roll supports itself; 2 mm-diameter roll does not
Very Plastic	2 mm-diameter roll supports itself

A1.2.3 Slope Description and Landscape Position

These interrelated aspects influence a site's suitability for installing DWMS by affecting the nature of surface water flow towards or away from the DWMS site. While the principles and terminology related to slope description and landscape position are more generally described in Section 3.3.5.3, they are presented in greater depth here so the reader can see how both aspects of the site affect the LTAR. Slope description relies on identifying both the downhill and lateral paths of the site's terrain as linear, concave or convex. Concave means that the site is located in an area similar to the bottom of a bowl, while the opposite shape (on top of an upside down bowl) is called convex. Within LTAR estimation, these terms are used to describe the nine slope type horizontal and vertical combinations (the first letter in each two letter classification referring to the downhill slope, and the second one referring to the lateral slope) as indicated in Figure A1.2.

The more concave the slope is, the lower the LTAR should be, because that implies a greater tendency for pooling of water within the site, which lowers a site's suitability for DWMS. This explains why generally speaking viewing from the top left to the bottom right of Figure A1.2, the LTAR adjustments become more negative (conservative). As with all of the sensorial soils tests or site topography-based observations described in this Appendix and in Section 3.3, the various classification systems and their associated LTAR adjustments (for slope description, in Table A1.7) are fully described later in the Appendix within the context of a full soils and site-based LTAR calculation case study.

While slope conditions may be thought of as more local, occurring at the specific site of the DWMS, landscape (slope) position may be thought of as more general, providing an assessment of the site relative to the greater environment as described in Section 3.3.5.3. Figure A1.3 is used in conjunction with Table A1.6 to determine the proper LTAR adjustment factor based on landscape position. During the site investigation, the service provider should determine the landscape position of the location chosen for the evaluation. Then he or she can simply match the position name with those provided for in Table A1.6.

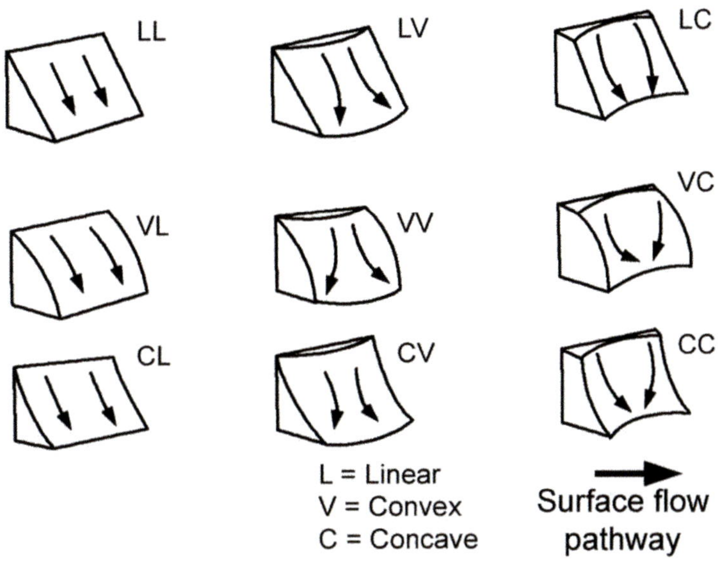

Figure A1.2 Slope description chart (Lindbo, 2009).

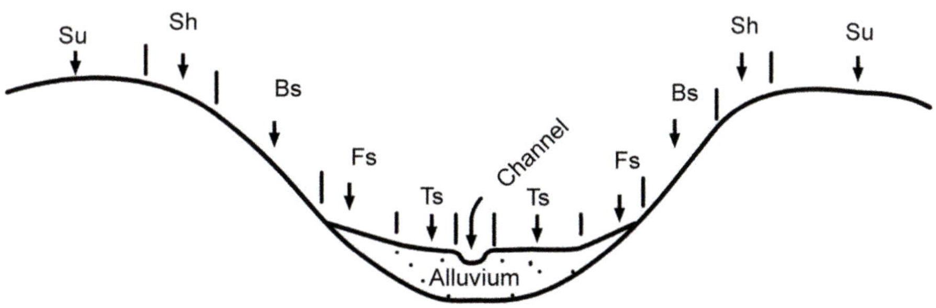

Figure A1.3 Landscape position slope names (Lindbo, 2009).

The slope names that link Figure A1.3 and Table A1.6 are:

Su – Summit or ridge	Bs – Back or side (lower)	Channel or drainage way
Sh – Shoulder	Fs – Foot slope	Depression
Bs – Back or side (upper)	Ts – Toe Slope	Floodplain

A1.3 STEPS FOR DETERMINING LTAR

(1) Use soil texture analysis to obtain the baseline LTAR value, which is the midpoint value in Table A1.1 based on grouping within soil groups I-IV, and add in any sub-texture adjustments within the relevant soil group.
(2) Make any other texture adjustments from Table A1.2.
(3) Identify the soil structure and apply those component adjustment factors as appropriate from Table A1.3.
(4) Determine the soil consistence and apply the most negative adjustment factor from those tests as found in Table A1.4.
(5) Determine any soil depth adjustment(s) as needed and apply the factor(s) from Table A1.5.
(6) Evaluate the landscape position and slope description and make any adjustments from Tables A1.6 and A1.7 as needed.

The sum of the baseline LTAR and adjustments gives the final LTAR. Be sure also correct the downslope trench edge depth (make it more shallow) for the slope if necessary using the slope correction factors (Table A1.8). The soils evaluation can then be completed by looking at landscape cover (vegetation, buildings, etc.) and microclimate conditions that may pose any additional challenges or opportunities for effluent dispersal at this site. Throughout the process, use the test hole soils evaluation log form (Figure A1.4) to record your results. Appendix II illustrates how each of these adjustments described above are made for a hypothetical site.

<table>
<tr><td colspan="9">Name</td></tr>
<tr><td colspan="9">Site</td></tr>
<tr><td colspan="9">Reference Number</td></tr>
<tr><td colspan="3">Date</td><td colspan="3">Time</td><td colspan="3">Weather</td></tr>
<tr><td colspan="9">Vegetation</td></tr>
<tr><td colspan="7">Landscape position and description</td><td colspan="2">Slope</td></tr>
<tr><td colspan="9">Aspect</td></tr>
<tr><td>Horizon</td><td>Depth (cm)</td><td>Texture Class</td><td>Matrix Color</td><td>Redoximorphic Feature Distinctness</td><td>Redoximorphic Feature Color</td><td>Structure (grade, size, type)</td><td>Consistence (moist or dry?) (failure, rupture, stickiness, plasticity)</td><td>Other</td></tr>
<tr><td></td><td></td><td></td><td></td><td></td><td></td><td></td><td></td><td></td></tr>
<tr><td></td><td></td><td></td><td></td><td></td><td></td><td></td><td></td><td></td></tr>
<tr><td></td><td></td><td></td><td></td><td></td><td></td><td></td><td></td><td></td></tr>
<tr><td></td><td></td><td></td><td></td><td></td><td></td><td></td><td></td><td></td></tr>
<tr><td></td><td></td><td></td><td></td><td></td><td></td><td></td><td></td><td></td></tr>
<tr><td></td><td></td><td></td><td></td><td></td><td></td><td></td><td></td><td></td></tr>
<tr><td></td><td></td><td></td><td></td><td></td><td></td><td></td><td></td><td></td></tr>
<tr><td colspan="9">Notes:</td></tr>
</table>

Figure A1.4 Test hole soils evaluation log form.

Appendix II

Example soils evaluation data interpretation and LTAR calculation

Using the data filled out in the sample test hole soils evaluation form (Figure A2.1) given below, the subsequent step-by-step process demonstrates how soils (and site) conditions can determine LTAR. It can also determine upslope and downslope trench wall depths, which are important because when the trenches run perpendicular to non-zero sloped-hills as is advised, their downslope edge must be slightly raised to compensate for the overall hill slope effect. The key soils conditions are texture, structure, and consistence, while the major site conditions are landscape position and slope description.

Landscape/slope position___Upper side slope, LL___Slope (%)___10___Shallowest limiting condition depth, inches (cm) ___ 36 (90)___Percentage of coarse fragments in restrictive horizon___10___

Horizon	Depth, inch (cm)	Texture (Group)	Structure (grade, size, type)	Consistence (moist, stickiness, plasticity)
1	0-6 (0-15)	Loam	weak, medium, granular	friable, slightly sticky, slightly plastic
2	6-20 (15-50)	Clay loam	moderate, medium, subangular blocky	friable, slightly sticky, slightly plastic
3	20-36 (50-90)	Clay	moderate, medium, subangular blocky	friable, slightly sticky, moderately plastic
4	36-45 (90-115)	Clay loam	moderate, coarse, subangular blocky	friable, slightly plastic, moderately plastic
5	45+ (115+)	Sandy loam	massive	friable, slightly sticky, slightly plastic

Figure A2.1 Example simplified filled-in test hole soils evaluation log form.

Example Soils Evaluation Procedure

Step 1: **Determining the depth for horizons and limiting condition.** The test hole analysis revealed that there were five horizons down to minimum advisable hole inspection depth of 1.2 m, with their depths measured and given in the proper column. While inspecting the horizons, clusters of mottles were seen at 36 in (90 cm) depth, which served as the shallowest limiting condition (SLC) and was marked as such in the form.

Step 2: **Determining the texture of each horizon, and the most-restrictive horizon overall.** Soil from each of the five horizons was tested using sensorial soils evaluation, which gave texture groups as listed above. Looking within the horizons, it was clear that horizon three, with clay was the most restrictive group, making horizon 3 the restrictive horizon (RH) and the starting point for the LTAR calculation. As shown in Table A1.1 the LTAR base was identified as 0.25 gallons per day per square foot (to be converted to metric units in the end), with no adjustment for sub-texture within the clay group. Also, information about the relative amount of coarse fragments in that horizon was used to make an additional LTAR adjustment, which in the end was not necessary, as shown in Table A1.2.

Step 3: **Determining the structure of each horizon.** The visual assessment and measuring processes were done as needed to evaluate the main structural aspects – ped grade, size and type – for each horizon and logged on the form. These observations were used for adjusting the LTAR by looking up (for the RH) each aspect's listing in Table A1.3 and adding the appropriate values to 0.25. The ped grade of moderate gave an adjustment of 0 (n/a); the medium size gave adjustment of 0; and the subangular blocky ped type gave an adjustment of 0. All of these three adjustments, when added to the LTAR from the previous step of 0.25, did not change that value.

It is important to note, however, that having three adjustment values of 0 is not guaranteed. Also, if any of the three aspects had been rated with LTAR adjustment values of 'UNS', that the soil would be classified as unstable (unsuitable) for dispersal (this is true for all soils with any parameters rated 'UNS' within the charts). In such cases, the RH should then be treated as the SLC, and a new RH above that should be selected, based on texture as described before. If sufficient soils depth (at least 75 cm) for dispersal is not present above the bottom of the new SLC, then another site with a sufficiently deep SLC and no unsuitable RH above that should be chosen.

Step 4: **Determining the consistence of each horizon.** The tests for resistance to breaking, stickiness, and plasticity were done for moist soil from each horizon, with findings recorded in the proper column of the log form. The results for horizon 3 (the RH) led to LTAR adjustments for those tests of 0 (friable), 0 (SS, or slightly sticky) and −0.05 (MP, moderately plastic) as determined by looking at Table A1.4. Given that the lowest value of all three should be chosen as the overall consistence LTAR adjustment, as described in that chart's footnote, the value of −0.05 was added to 0.25 to get a revised LTAR of 0.20.

Step 5: **Determining the soil depth between the bottom of the dispersal system and the SLC.** Given the SLC depth of 36 inches (90 cm), the need for a soil buffer of 12 inches (30 cm) above that, and the fact that as mentioned earlier at least 18 inches of trench basin/soil cover must be above the trench bottom, this limited the trench bottom depth to a range of 18 inches to 24 inches (36 in minus 12 in). The 18 inches (45 cm) value was chosen in order to maximize the distance to the SLC (18 inches), which still falls within the 12–18 inch range as specified in Table A1.5. This gave an LTAR adjustment of −0.05 (for Group II–IV texture soils, which as shown in Table A1.1 are everything but sand and loamy sand, which is Group I). This led to a revised LTAR of 0.15. However, if that distance to SLC could have been greater than 18 inches – meaning SLC was below 90 cm depth – then as shown in Table A1.5, an 'n/a' or even +0.05 LTAR adjustment might have been possible, which shows quantitatively how choosing better, deeper soil sites allows increased dispersion rates.

Step 6: **Determining the site's landscape position and slope description.** With the help of experienced surveyors, it was found that the chosen dispersal site had an upper side slope landscape position and linear, linear (LL) overall slope. Tables A1.6 and A1.7 were used to conclude that the two LTAR adjustment factors for those two findings were −0.05 and 0, which added to the LTAR gave a final LTAR of 0.10 g/d/ft^2 (4.07 L/d/m^2).

As for other impacts on soils dispersal design, if the site slope is greater than or equal to 5 percent (not ideal), the downslope trench downslope wall bottom depth needs to be raised with additional soil covering based on an adjustment value as shown in Table A1.8. In this case, the slope of 10 percent implies an adjustment of 4 in, equal to 8 cm (all based on an implied maximum acceptable trench width of 36 inches, or 90 cm).